Lecture Notes in Control and Information Sciences

Edited by M. Thoma and A. Wyner

Vol. 97: I. Lasiecka/R. Triggiani (Eds.)
Control Problems for Systems
Described by Partial Differential Equations
and Applications
Proceedings of the IFIP-WG 7.2
Working Conference
Gainesville, Florida, February 3-6, 1986
VIII, 400 pages, 1987.

Vol. 98: A. Aloneftis
Stochastic Adaptive Control
Results and Simulation
XII, 120 pages, 1987.

Vol. 99: S. P. Bhattacharyya
Robust Stabilization Against
Structured Perturbations
IX, 172 pages, 1987.

Vol. 100: J. P. Zolésio (Editor)
Boundary Control and Boundary Variations
Proceedings of the IFIP WG 7.2 Conference
Nice, France, June 10-13, 1987
IV, 398 pages, 1988.

Vol. 101: P. E. Crouch,
A. J. van der Schaft
Variational and Hamiltonian
Control Systems
IV, 121 pages, 1987.

Vol. 102: F. Kappel, K. Kunisch,
W. Schappacher (Eds.)
Distributed Parameter Systems
Proceedings of the 3rd International Conference
Vorau, Styria, July 6–12, 1986
VII, 343 pages, 1987.

Vol. 103: P. Varaiya, A. B. Kurzhanski (Eds.)
Discrete Event Systems:
Models and Applications
IIASA Conference
Sopron, Hungary, August 3-7, 1987
IX, 282 pages, 1988.

Vol. 104: J. S. Freudenberg/D. P. Looze
Frequency Domain Properties of Scalar
and Multivariable Feedback Systems
VIII, 281 pages, 1988.

Vol. 105: Ch. I. Byrnes/A. Kurzhanski (Eds.)
Modelling and Adaptive Control
Proceedings of the IIASA Conference
Sopron, Hungary, July 1986
V, 379 pages, 1988.

Vol. 106: R. R. Mohler (Editor)
Nonlinear Time Series and
Signal Processing
V, 143 pages, 1988.

Vol. 107: Y. T. Tsay, L.-S. Shieh, St. Barnett
Structural Analysis and Design
of Multivariable Systems
An Algebraic Approach
VIII, 208 pages, 1988.

Vol. 108: K. J. Reinschke
Multivariable Control
A Graph-theoretic Approach
274 pages, 1988.

Vol. 109: M. Vukobratović/R. Stojić
Modern Aircraft Flight Control
VI, 288 pages, 1988.

Vol. 110: B. J. Daiuto, T. T. Hartley,
S. P. Chicatelli
The Hyperbolic Map and Applications
to the Linear Quadratic Regulator
VI, 114 pages, 1989

Vol. 111: A. Bensoussan, J. L. Lions (Eds.)
Analysis and Optimization
of Systems
XIV, 1175 pages, 1988.

Vol. 112: Vojislav Kecman
State-Space Models of Lumped
and Distributed Systems
IX, 280 pages, 1988

Vol. 113: M. Iri, K. Yajima (Eds.)
System Modelling and Optimization
Proceedings of the 13th IFIP Conference
Tokyo, Japan, Aug. 31 – Sept. 4, 1987
IX, 787 pages, 1988.

Vol. 114: A. Bermudez (Editor)
Control of Partial Differential Equations
Proceedings of the IFIP WG 7.2
Working Conference
Santiago de Compostela, Spain, July 6–9, 1987
IX, 318 pages, 1989

Vol. 115: H.J. Zwart
Geometric Theory for Infinite
Dimensional Systems
VIII, 156 pages, 1989.

Vol. 116: M.D. Mesarovic, Y. Takahara
Abstract Systems Theory
VIII, 439 pages, 1989

For information about Vols. 1-96 please contact your bookseller or Springer-Verlag

Lecture Notes in Control and Information Sciences
Edited by M. Thoma and A. Wyner

171

P. Stefanidis, A. P. Papliński, M. J. Gibbard

Numerical Operations with Polynomial Matrices

Application to Multi-Variable Dynamic Compensator Design

Springer-Verlag
Berlin Heidelberg GmbH

Authors
Peter Stefanidis
Communica Software Consultants
Level 1
170 Greenhill Rd.
Parkside 5063
Australia

Andrzej Piotr Papliński
Dept. of Robotics and Digital Technology
Monash University
P. O. Box 197
Caulfield East 3145
Victoria
Australia

Michael John Gibbard
Dept. of Electrical and Electronic Engineering
The University of Adelaide
GPO Box 498
Adelaide 5001
Australia

ISBN 978-3-540-54992-5 ISBN 978-3-540-46633-8 (eBook)
DOI 10.1007/978-3-540-46633-8

Originally published by Springer-Verlag Berlin Heidelberg New York in 1992

Typesetting: Camera ready by authors

61/3020-5 4 3 2 1 0 Printed on acid-free paper.

Acknowledgements

Michael Gibbard is, and Andrew Papliński and Peter Stefanidis were, in the Department of Electrical and Electronic Engineering at The University of Adelaide, South Australia.

We gratefully acknowledge the financial support provided by both a Commonwealth Postgraduate Research Award for Peter Stefanidis and a grant from the Defence Science and Technology Organisation, Department of Defence, Australia for Andrew Papliński.

Contents

List of Figures

Chapter 1

Introduction

1.1 Motivation

The objective at the outset of this project was to devise a *direct* method for the design of dynamic compensators for multi–input multi–output (MIMO) control systems. For the systems of concern,

(i) only a limited number of signals, including the output variables, was available for feedback,

(ii) the specifications for the dynamic performance of the compensated system required that its poles lie within a defined region of the complex s– or z–plane,

(iii) the steady-state performance of the compensated system was to be specified in terms of the alignment or following errors between the output and reference (set–point) signals,

(iv) the design technique was to be applicable to both continuous and digital controllers.

Since the number of signals available for feedback was limited, the compensator may be of order one or greater, i.e. it possesses dynamics and was thus called a 'dynamic' compensator. By a 'direct' design method it was implied that the compensator parameters should be calculated directly from the steady–state and transient performance specifications. Because the transient performance specifications are often cast in terms of closed–loop pole positions, it seemed that a pole–assignment technique would not only be an appropriate design approach but might lend itself to implemention as a direct design method for MIMO systems. Some of the relevant background material on pole–assignment methods is covered in the next section.

A seemingly natural approach for the implementation of pole–assignment methods is one based on a transfer function description of the plant and of the open– and closed–loop systems. For single–input single–output (SISO) systems, computations involving transfer functions are relatively straightforward, however, this is not the case for MIMO systems. In particular, for the purposes of this project, suitable computational methods for calculations involving MIMO transfer function representations did not appear to be available. Consequently it was necessary to select an appropriate representation for the MIMO system and to develop a new computational methodology to perform the calculations required in the proposed direct design method. For the MIMO transfer function representation it was decided to employ the so–called Matrix Fraction Description (MFD), in which the numerator and denominator terms are polynomial matrices. For fast and efficient computatation it was deemed desirable that all calculations involve matrices of real numbers rather than operations on polynomial matrices. The basis for such a computational methodology is explained in Section 1.3.

The motivation for this monograph is that, through publication, we can draw attention to the numerical representation of polynomial matrices, and applications in computation and analysis of MIMO systems. For computational purposes, the numerical representation – by matrices of real numbers – is readily implemented in software packages such as $Matrix_x$ [53] and *Matlab* [73]. Four examples of compensator design are included, not only to illustrate the computational methods but also to attract the interest of the design engineer to whom the significance and application of algebraic methods may otherwise appear somewhat abstract. Furthermore, we hope that this monograph may provide some tools and incentive to other workers to improve computational and design methods for MIMO control systems.

1.2 Pole Assignment

Multivariable compensator design has been a fertile research area for many years. In particular, the problem of pole assignment, which has been studied thoroughly [14,20, 22,23,29,37,38,45,82,89,91,95,96,103,108], is concerned with placing, by means of state or output feedback, the poles of a closed-loop, linear time-invariant system in specified locations in the s– or z–plane (for continuous– or discrete–time systems, respectively). The state–feedback approach is well established; however, the output–feedback case has been a continuing field of research [114,117]. Because it is not possible to measure all of the system states in many industrial applications, we shall concern ourselves only with the more practical case of output feedback.

A number of methodologies exists for the solution to the pole–assignment problem. Let us consider briefly some well–known design methods based on the state-space and transfer function matrix descriptions of dynamic systems. These methodologies can be distinguished by the type of approach. Both the state–feedback and output–feedback methods result in compensator matrices which are, broadly speaking, either dyadic, i.e. have rank equal to one, or have full rank.

An approach to this problem, which was developed by Kalman and Bucy [60], and by Luenberger [71,72], is to estimate the system states if they are not available from the measured variables. Two techniques for constructing estimates of the states are known as the Kalman–Bucy filter and the Luenberger observer. Having a method for estimating the states, the designer is then free to select one of the many well–developed state–feedback techniques to implement pole assignment.

The methodology used in the design of dyadic compensators generally begins with the simplification of a least–order, MIMO plant model to a least–order, single–input multi–output plant model, with the feedback compensator then formed by the outer product of two vectors. The problem of solving for the feedback compensator is then one of equating the coefficients of the characteristic equation of the resulting closed–loop system and solving a set of linear equations. A methodology described by Munro and Novin-

Hirobod [83] constructs a full rank static or dynamic compensator from a sequence of dyads. Although the dyadic algorithms have considerable simplicity, the resulting closed-loop systems have poor disturbance rejection properties compared with their full rank counterparts [12].

In this publication, an approach based on algebraic theory is employed for the design of an output–feedback compensator using pole assignment. In algebraic theory a *fractional representation* of a linear system is adopted. The central feature of this approach is to factor the plant $G(q)$ and feedback compensator $F(q)$ as a product of a polynomial matrix and the inverse of second polynomial matrix; i.e., $G(q) = N_{\mathrm{R}}(q).D_{\mathrm{R}}^{-1}(q)$ and $F(q) = V^{-1}(q).K(q)$. These forms are known as matrix fraction descriptions. There exists an extensive body of literature on the applications of matrix fraction descriptions and, in particular, their use in the design of linear, time–invariant, MIMO control systems [7] [16] [50] [57] [92] [111] [114].

In later chapters, the design of a two–element compensator [114] shown in Fig. 1.1 is discussed. This type of compensator enables the designer to satisfy the steady–state performance specifications through the pre–compensator, $R(q)$, and the dynamic performance specifications through the output–feedback compensator, $F(q)$. Let us consider the design of the latter element first.

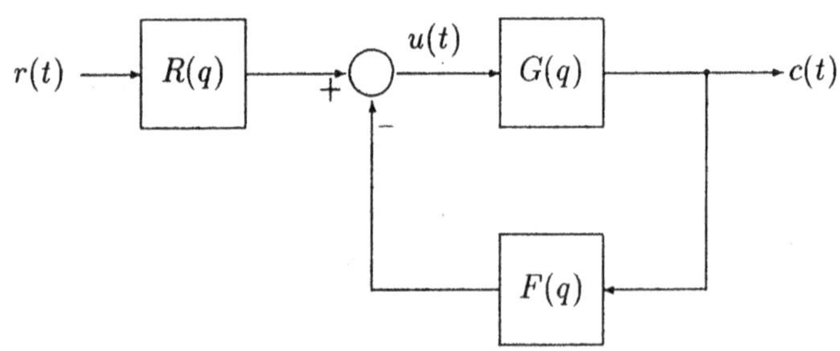

Figure 1.1: Two–element compensation scheme

The central problem in designing dynamic compensation by means of algebraic methods is to find a coprime, proper (causal), least–order feedback compensator $F(q)$ by solving a polynomial matrix equation (the so-called left compensator equation),

$$W(q) = V(q).D_{\mathrm{R}}(q) + K(q).N_{\mathrm{R}}(q)\,. \tag{1.1}$$

In this equation the polynomial matrix, $W(q)$, contains information on the designer—assigned and the solution-assigned closed-loop pole locations. Accordingly, based on pole assignment, we shall present a methodology for solving the left compensator equation and realising a continuous or discrete output-feedback compensator.

The method of solution of an equation of the form of eqn. (1.1) is well-developed [16,17,39,118]. However, we shall modify the form of the equation so that the problem of determining the compensator parameters is cast as a minimal design problem [58]. This methodology searches for a minimal basis of the left null space of the left compensator equation; a solution for the parameters of the feedback element, $F(q)$, of the compensator can then be extracted from the minimal basis.

From the steady-state performance specifications for the design, the pre-compensator element, $R(q)$, (a polynomial matrix) is determined. In the design methodology, the final value theorem is used to realise the steady-state specifications for the closed-loop system in terms of either the alignment or following errors.

The design of a compensator must not only implement pole placement and the desired steady-state performance, but it also requires that the performance of the closed-loop system is insensitive to plant or compensator parameter variations, external disturbances, etc. While these factors will not be discussed in the following chapters, they should also be included in a practical design procedure.

1.3 Computational Aspects of the Design

In the following chapters a new class of computational methods and the associated methodology for the design of digital or continuous-time compensators are investigated and developed. These compensators are for use in linear, time-invariant MIMO systems. The computational methods are based on the algebraic properties of such systems. In addition, new direct methods have been formulated for:

- computing the determinant and inverse of a non-singular polynomial matrix;
- constructing a non-singular polynomial matrix, the eigenvalues of which are specified by the designer;
- reducing a full rank polynomial matrix to its row-proper form;
- performing transformations between the various models of a linear system; and
- designing the dynamic and steady-state compensation of a linear, time-invariant, MIMO system.

Furthermore, results in support of the above methods also include:

- efficient methods for carrying out operations on polynomial matrices.

  [Systems, which are represented by polynomial matrix models, are described by a numerical representation of such matrices; i.e., a polynomial matrix $L(q)$ will be represented by a row block-matrix L of polynomial coefficients. Thus calculations involving a polynomial matrix $L(q)$ become operations on a row block-matrix L. The advantages in operating on the block-matrix representation are that

  - matrix operations are easily implemented in computer software,
  - with matrices of real numbers, conventional matrix algebra can be used in analysis.]

- the Toeplitz-Sylvester block-matrix structure, known as the *resultant*, with which the product of two polynomial matrices can be formed. (The resultant has also been called a generalised resultant matrix [65].)

  [Such a product can be expressed as the matrix product of the row block-matrix of coefficients of a multiplicand polynomial matrix and the resultant of a multiplier polynomial matrix. This expression can be considered as a closed form of the well-known iterative algorithms for the multiplication of polynomials. The resultant of a polynomial matrix will be employed in a number of the computational techniques developed for the design process.]

- the concepts of the minimal polynomial basis, the "structure" matrix and its numerical representation known as the "skeleton" block-matrix.

It should be emphasized that the algorithms have been developed in this project are purely functional, that is, no attempt has been made at this stage to make them more efficient numerically, etc.

The computational techniques, discussed above, are implemented in a commercially available software package [53], which is a powerful, programmable matrix calculator with graphics capabilities. The numerical techniques developed for the design procedures are combined with the facilities available in the commercial package to create the software package MIMOCOD. The latter package minimises the participation by the designer in the computations at each step of the design process and, further, enables the designer to determine whether the closed-loop specifications are realised through time-response displays, etc.

1.4 Layout

Chapter 2 is devoted to defining the properties of a polynomial matrix and its associated numerical representation. The concepts of a minimal polynomial basis, the "structure" matrix and its numerical description known as the "skeleton" block-matrix are defined. Various methodologies which are required in the design process are also discussed and developed.

In Chapter 3, methods of transforming between the various descriptions of a linear, time-invariant, MIMO system are presented. These descriptions are the state-space model, the matrix fraction description and the transfer function matrix representation. The transformations are based on the ideas developed in Chapter 2. The AutoRegressive-Moving Average (ARMA) model of a system transfer function is also considered; this model is used for the construction of a control algorithm for a digital controller.

In Chapter 4, based on the fractional approach, a method for designing dynamic compensators for MIMO systems is presented. The design of a two-element compensator is examined, for which both the dynamic and steady–state responses of the closed–loop system are considered. The dynamics of the closed–loop system are altered through pole placement, and steady–state design is based on alignment– and following–error specifications.

In Chapter 5, four design examples are presented; a description of the design steps and the results realised at various design steps are provided. These steps emulate those that a designer would follow using the sofware package MIMOCOD to design a two–element compensator. In Chapter 6 some conclusions are drawn on the results in the earlier chapters and a number of suggestions is made for future work.

Chapter 2

Polynomial Matrices and Related Operations

2.1 Introduction

In this section we shall consider polynomial matrices and their numerical representation, which we have called the row/column block–matrix description, and investigate their properties. Algebraic operations, which employ the block–matrix description of polynomial matrices, and the associated computational techniques will be examined.

2.2 Properties of Polynomial Matrices

Definition 2.1 A polynomial matrix $L(q)$ is a rectangular array of n_{r} rows and n_{c} columns with elements from the ring of polynomials in q, $\Re[q]$, with coefficients in the field of real numbers, $\Re$. An element of $L(q)$ is

$$\ell_{ij}(q) = l_{ij}.Q_{\partial[\ell_{ij}(q)]},\ i \in \underline{n}_{\mathrm{r}},\ j \in \underline{n}_{\mathrm{c}},$$

where

$$l_{ij} = \left[\begin{array}{cccc} l_{0_{ij}} & l_{1_{ij}} & \dots & l_{\partial[\ell_{ij}(q)]_{ij}} \end{array}\right]$$

is a row vector of the real coefficients of the ith row, jth column polynomial element $\ell_{ij}(q) \in \Re[q]$ of degree $\partial[\ell_{ij}(q)]$, and

$$Q_k = [q^k \dots 1]^T$$

is a column vector of the powers in some indeterminate q (e.g., $q = s,\ z,\ D$) [7,8,46].

$\nabla\nabla\nabla$

If the elements of a matrix $\hat{L}$ belong to the field $\Re$, then the operations applicable to $\hat{L}$ also apply to $L(q)$ with one exception: for the existence $L^{-1}(q) = \mathrm{Adj}(L(q))/\det(L(q))$, of a non-singular polynomial matrix $L(q) \in \Re[q]^{n\times n}$, it is not sufficient to have $\det(L(q)) \neq 0$. The determinant is an element of $\Re[q]$, and for $L^{-1}(q)$ to exist $\det(L(q))$ must be a unit in $\Re[q]$ [102]. In other words, $\det(L(q))$ must be a non-zero scalar for the inverse of $L(q)$ to be a matrix the elements of which are polynomials – rather than rational polynomials – in q.

Definition 2.2 A polynomial matrix $L(q) \in \Re[q]^{n_r \times n_c}$ can be written

$$\begin{aligned} L(q) &= L_0 q^m + L_1 q^{m-1} + \cdots + L_{m-1} q^1 + L_m \qquad (2.1) \\ &= \begin{bmatrix} L_0 & L_1 & \dots & L_{m-1} & L_m \end{bmatrix} . \left(\begin{bmatrix} q^m \\ q^{m-1} \\ \vdots \\ q^1 \\ 1 \end{bmatrix} \otimes I_{n_c} \right) \\ &= L . [Q_m \otimes I_{n_c}], \end{aligned}$$

where:

- the *degree* $m = \partial[L(q)]$ of $L(q)$ is the highest degree of the indeterminate of $L(q)$; the associated (leading) coefficient block-matrix is $L_0 \not\equiv 0_{n_r \times n_c}$;
- $L_j \in \Re^{n_r \times n_c}$ is a block-matrix consisting of the coefficients of the indeterminate of degree $m - j$ of the polynomial elements of $L(q)$;
- $L \in \Re^{n_r \times n_c . (m+1)}$ is a matrix of coefficient block-matrices of $L(q)$;

- I_{n_c} is the $n_c \times n_c$ identity matrix, and $\otimes$ denotes the Kronecker product.

The form of the polynomial matrix described by eqn. (2.1) will be used to define the numerical representation of a polynomial matrix.

2.2.1 Minimal Polynomial Bases

In linear system theory one frequently deals with matrices the elements of which are polynomials in some indeterminate q with coefficients in some field $\mathcal{F}$ (e.g., the field of real or complex numbers). The set of polynomials in q over $\mathcal{F}$ forms a ring, conventionally denoted $\mathcal{F}[q]$.

If $L(q)$ is a $n_r \times n_c$ polynomial matrix with elements in some ring, its column space (the set of all linear combinations of its columns) is a polynomial vector space $\mathcal{F}[q]$ over that ring, and the dimension of $\mathcal{F}[q]$ is the rank of $L(q)$. Conversely, if $\mathcal{F}[q]$ is a polynomial vector space of n_r-tuples of ring elements of dimension n_c, it has a basis of n_c linearly independent n_r-tuples, which may be arranged to form a $n_r \times n_c$ polynomial matrix $L(q)$ of rank n_c. Any $n_r \times n_c$ polynomial matrix which has rank n_c is called a full–rank polynomial matrix (implying $n_c \leq n_r$) and forms a basis for the corresponding vector space.

From linear algebra we know that two full–rank $n_r \times n_c$ polynomial matrices $L_1(q)$ and $L_2(q)$ are bases for the same vector space if and only if $L_1(q) = L_2(q).T(q)$ for some full–rank, non–singular, $n_c \times n_c$ polynomial matrix $T(q)$ [4]. Our objective is to consider certain canonical bases for the vector space of n_r-tuples over the ring of polynomials $\mathcal{F}[q]$.

Let us introduce the following definitions and a theorem found in Forney [40], Wolovich [118] and Patel [92] which will be used extensively throughout Chapters 2, 3 and 4.

Proposition 2.1 Let $L(q)$ be a $n_r \times n_c$ polynomial matrix with elements in the ring of polynomials $\Re[q]$, where $n_r \geq n_c$ (see Remark 2.2 for $n_r < n_c$).

Definition 2.3 [40]: The *column degree* $\partial_{cj}[L(q)]$ of the jth column n_r-tuple $[\ell_{1j}(q) \ldots \ell_{n_r j}]^T$ of polynomials in $L(q)$ is the greatest degree of its jth column polynomial elements $\ell_{ij}(q)$, $i \in \underline{n}_r$.

Definition 2.4 [40]: The jth *column index* of $L(q)$ is defined as $\nu_j = \partial_{cj}[L(q)]$, $j \in \underline{n}_c$, and the *order* of $L(q)$ is defined as $\nu = \sum_{j=1}^{n_c} \nu_j$.

Definition 2.5 [40]: Assume $\mathcal{F}[q]$ is a n_c-dimensional column vector space of n_r-tuples with elements over $\Re[q]$. A *minimal basis* of $\mathcal{F}[q]$ is a polynomial matrix $L(q) \in \Re[q]^{n_r \times n_c}$ such that $L(q)$ is a basis for $\mathcal{F}[q]$; moreover $L(q)$ has *least* order among all polynomial bases for $\mathcal{F}[q]$.

$\nabla\nabla\nabla$

We now consider a simple lower bound on the order of a minimal basis. The minors of dimension $n_c \times n_c$ of a $n_r \times n_c$ polynomial matrix $L(q)$ are the determinants of all $C_{n_c}^{n_r}$ square $n_c \times n_c$ submatrices obtained by selecting subsets of n_c rows of $L(q)$. ($C_{n_c}^{n_r}$ denotes the number of different combinations of n_r different rows, n_c at a time, without repetition). The minors, of dimension $n_c \times n_c$, of $L(q).T(q)$ are those of $L(q)$ multiplied by $\det(T(q))$. Hence all bases of a given polynomial vector space over $\mathcal{F}[q]$ have the same set of minors, of dimension $n_c \times n_c$, up to a common multiple. Let us define the *normalised minors* of $L(q)$ as its minors, of dimension $n_c \times n_c$, divided by their greatest common divisor. The maximum degree of any minor, of dimension $n_c \times n_c$, cannot exceed the order of $L(q)$, since a minor is a sum of products of polynomials, one from each column. Furthermore, this maximum degree cannot be less than the maximum degree of the normalised minors of $L(q)$. Hence the order of a minimal basis for $\mathcal{F}[q]$ is lower–bounded by the maximum degree of the normalised minors of $L(q)$ [40].

Remark 2.1 A minimal basis $L(q) \in \Re[q]^{n_r \times n_c}$ is considered as a canonical basis for the column vector space of n_r-tuples over $\Re[q]$ [40,92]. In general, a polynomial vector space $\mathcal{F}[q]$ has many minimal bases. For some purposes it may be desirable to specify one of these bases as canonical, so that the polynomial vector space can be uniquely identified with its canonical basis.

Definition 2.6 [40,118]: The highest column–degree coefficient matrix $\Gamma_c[L(q)]$ of $L(q)$ is the $n_r \times n_c$ $\Re$-matrix, the jth column of which consists of the coefficients of q^{ν_j} in the jth column of $L(q)$.

Theorem 2.1 [40]: Let $\mathcal{F}[q]$ be a n_c-dimensional column vector space of n_r-tuples over $\Re[q]$, and let $L(q)$ be $n_r \times n_c$ polynomial basis for $\mathcal{F}[q]$ with indices ν_j, $j \in \underline{n_c}$, and order $\nu = \sum_{j=1}^{n_c} \nu_j$. Then the following statements are equivalent:

1. $L(q)$ is a minimal basis for $\mathcal{F}[q]$.

2. The highest column-degree coefficient matrix $\Gamma_c[L(q)]$ is of full column rank; i.e., $\text{rank}(\Gamma_c[L(q)]) = n_c$.

3. (a) The greatest common divisor of minors of dimension $n_c \times n_c$ of $L(q)$ is 1, and

   (b) their greatest degree is ν.

Result 2.1 [40]: Item 2 of Theorem 2.1 is true if and only if item 3b of Theorem 2.1 is true. Furthermore, the coefficient of q^ν in a given $n_c \times n_c$ minor of $L(q)$ is the corresponding $n_c \times n_c$ minor of $\Gamma_c[L(q)]$.

Definition 2.7 [92,118]: Let the polynomial matrix $L(q) \in \Re[q]^{n_r \times n_c}$ be a minimal basis (i.e., of least order) for a n_c-dimensional column vector space of polynomial n_r-tuples over $\mathcal{F}[q]$. Then $L(q)$ is said to be *column-proper*.

Note the highest column-degree coefficient matrix, $\Gamma_c[L(q)]$, of $L(q)$ is full-rank. Furthermore, $L(q)$ is said to be *row-proper* if and only if its transpose is column-proper.

Remark 2.2 It is clear that a similar theorem and similar definitions can be realised for a n_r-dimensional row vector space of n_c-tuples, where $n_r \leq n_c$. Moreover, the row vector space is related to the column vector space through its transpose. The terms "column", "ν" and "$\Gamma_c[L(q)]$" should be then replaced by "row", "μ" and "$\Gamma_r[L(q)]$", respectively, when considering the row vector space.

Let us illustrate the above concepts by an example.

Example 2.1 Consider the 5×3 polynomial matrix

$$L(q) = \begin{bmatrix} q+2 & 0.5 & q^2+1.5 \\ 1 & q-0.5 & 1.5 \\ 1 & -0.5 & 2q^3+1.5q+1 \\ 1 & 6q+2.5 & 0.5 \\ 3q+2 & -0.5 & 0.5 \end{bmatrix}.$$

The column indices of $L(q)$ are $\nu_1 = 1$, $\nu_2 = 1$, $\nu_3 = 3$ with the order $\nu = \sum_{j=1}^{3} \nu_j = 5$. The highest column-degree coefficient matrix $\Gamma_c[L(q)]$ is

$$\Gamma_c[L(q)] = \begin{bmatrix} 1 & 0 & 0 \\ 0 & 1 & 0 \\ 0 & 0 & 2 \\ 0 & 6 & 0 \\ 3 & 0 & 0 \end{bmatrix}$$

from which it is clear that $\text{rank}(\Gamma_c[L(q)]) = 3$ implying that $L(q)$ is column-proper as well as being a minimal basis for the column space it spans.

2.2.2 Properties of a Square Polynomial Matrix

Proposition 2.2 Let $L(q)$ be a $n \times n$ non-singular polynomial matrix with elements in the ring of polynomials $\Re[q]$ and degree $m = \partial[L(q)]$. Then we can state the following results.

Definition 2.8 The determinant of $L(q)$ is a polynomial $l(q) \in \Re[q]$. The *polynomial order* of $l(q)$ is $\partial[l(q)]$. The eigenvalues of $L(q)$ are the zeros of $l(q)$, that is, all $\lambda \in C$ where $\det(L(\lambda)) = 0$.

Definition 2.9 [45,8]: The polynomial matrix $L(q)$ is said to be:

- *regular* if the leading coefficient block-matrix L_0 is non-singular;
- *monic* if the leading coefficient block-matrix L_0 is the identity matrix; i.e., $L_0 = I_n$;
- *co-regular* if the trailing coefficient block matrix L_m is non-singular; and
- *unimodular* if its determinant is a non-zero scalar in $\Re$. In this case, the inverse $L^{-1}(q)$ is also a polynomial matrix.

Remark 2.3 We should note, with reference to item 3b of Theorem 2.1, that if $L(q)$ is column-proper (row-proper) then $\nu = \partial[l(q)] = \sum_{j=1}^{n} \nu_j$ ($\mu = \partial[l(q)] = \sum_{i=1}^{n} \mu_i$). In other words, the polynomial order of $\det(L(q))$ is equivalent to the order of the polynomial matrix $L(q)$; i.e., $\partial[l(q)] = \nu$.

Result 2.2 [8]: An important inequality exists that relates the dimension, n, and degree, m, of $L(q)$, and the polynomial order, $\partial[l(q)]$, of its determinant, namely

$$\partial[l(q)] \leq n.m\,. \tag{2.2}$$

In other words, the number of eigenvalues of $L(q)$ (i.e. the polynomial order of $l(q)$) cannot exceed the product of the dimension and degree of the polynomial matrix. The equality holds when the polynomial matrix is regular or monic.

2.3 Numerical Representation of Polynomial Matrices

The form of the non-square polynomial matrix $L(q)$ described by eqn. (2.1) of Definition 2.2 is unsuitable for the development of computational algorithms used in computer software. A representation of the polynomial matrix, which embodies the structural properties of the polynomial matrix and is capable of being easily implemented in computer algorithms, is known as the block-matrix representation of the polynomial matrix [45,7]. The polynomial matrix can be expressed in the form described by eqn. (2.1), in which *the block-matrix description consists solely of the coefficients of the polynomial matrix.* The block-matrix can be expressed in either row block-matrix or column block-matrix form. If we consider the polynomial matrix $L(q)$ of eqn. (2.1) and its two equivalent block-matrix descriptions, the relationship between the three descriptions can be clearly seen. Firstly, the block-matrix of coefficients

$$L = \begin{bmatrix} L_0 & L_1 & \ldots & L_m \end{bmatrix}, \tag{2.3}$$

is defined as the *row* block-matrix representation of the polynomial matrix $L(q)$, because each of the coefficient block-matrices are aligned to form a *row* of coefficient block-matrices. Secondly, the block-matrix of coefficients

$$L^b = \begin{bmatrix} L_0 \\ L_1 \\ \vdots \\ L_m \end{bmatrix}, \tag{2.4}$$

for a similar reason, is defined as the *column* block–matrix representation.

The coefficient block–matrices L_m and L_0 correspond to the zero and mth degree indeterminates, respectively, of $L(q)$ (see Definition 2.2).

2.4 Operations on the Numerical Representation of a Polynomial Matrix

2.4.1 Introduction

The row or column block–matrix descriptions are alternative numerical representations of a polynomial matrix. For simplicity, the algorithms in the following sections will employ the row block–matrix description of a polynomial matrix.

In this section we examine:

- the operations of addition, subtraction, block–transposition and multiplication involving row block–matrices;
- the construction of the resultant of a block–matrix used in the multiplication of polynomial matrices;
- the properties of a polynomial matrix known as the "structure" matrix and its equivalent block–matrix description known as the "skeleton" block–matrix;
- an algorithm which constructs a non–singular polynomial matrix with specified properties;
- an algorithm which computes the determinant of a square polynomial matrix; and
- an algorithm which transforms a full row–rank polynomial matrix to its row–proper form.

2.4.2 Addition or Subtraction of Row Block–Matrices

The addition of block–matrices follows the rules of addition for ordinary matrices. Row block–matrix addition involves the addition of corresponding elements of the row block–matrices. Only row block–matrices of two polynomial matrices of the same dimension can be summed. The column dimension of the row block–matrix is dependent on both the *degree* and *column dimension* of its polynomial matrix. In the case of the row block–matrices satisfying the dimension criterion, but having different degrees, the block–matrix which has the smaller degree of the two must be padded with null block–matrices. This increases the column dimension of the padded row block–matrix to that of the dimension of the higher degree row block–matrix. Addition of the two row block–matrices can then proceed. Similarly, the above rules apply to the subtraction of row block–matrices.

Example 2.2 Consider the following two 3×2 polynomial matrices

$$C(q) = \begin{bmatrix} q^3+3q-2 & q-1 \\ q^2+6q-4 & 10 \\ q^3+2q^2-3q+1 & 5q-8 \end{bmatrix}, \text{ and } D(q) = \begin{bmatrix} q+3 & q-4 \\ 3q & 2q-1 \\ 10 & q-2 \end{bmatrix},$$

the degrees of which are $\partial[C(q)] = 3$ and $\partial[D(q)] = 1$, respectively.

The row block–matrix representations of $C(q)$ and $D(q)$ are

$$C = \left[\begin{array}{cc|cc|cc|cc} 1 & 0 & 0 & 0 & 3 & 1 & -2 & -1 \\ 0 & 0 & 1 & 0 & 6 & 0 & -4 & 10 \\ 1 & 0 & 2 & 0 & -3 & 5 & 1 & -8 \end{array}\right], \text{ and } D = \left[\begin{array}{cc|cc} 1 & 1 & 3 & -4 \\ 3 & 2 & 0 & -1 \\ 0 & 1 & 10 & -2 \end{array}\right],$$

respectively.

Before the row block–matrices C and D can be summed, the row block–matrix D is padded with two null block–matrices, resulting in the column dimension of D being equal to that of C.

Hence

$$D = \left[\begin{array}{cc|cc|cc|cc} 0 & 0 & 0 & 0 & 1 & 1 & 3 & -4 \\ 0 & 0 & 0 & 0 & 3 & 2 & 0 & -1 \\ 0 & 0 & 0 & 0 & 0 & 1 & 10 & -2 \end{array}\right].$$

Summation of the row block–matrices C and D yields

$$C + D = \left[\begin{array}{cc|cc|cc|cc} 1 & 0 & 0 & 0 & 4 & 2 & 1 & -5 \\ 0 & 0 & 1 & 0 & 9 & 2 & -4 & 9 \\ 1 & 0 & 2 & 0 & -3 & 6 & 11 & -10 \end{array}\right],$$

or, equivalently,

$$C(q)+D(q)=\begin{bmatrix} q^3+4q+1 & 2q-5 \\ q^2+9q-4 & 2q+9 \\ q^3+2q^2-3q+11 & 6q-10 \end{bmatrix}.$$

2.4.3 Block–Transpose of a Block–Matrix

The block–transpose operation provides a link between a row block-matrix description and its column block-matrix description, and visa-versa. The block–transpose operation applied to a matrix of block–matrices is similar to the transpose operation applied to an ordinary matrix. We shall find shortly a need to determine the block–transpose of a matrix of block–matrices having the form

$$M=\begin{bmatrix} M_0 & \dots & \dots & M_m & 0 & \dots & \dots & 0 \\ 0 & M_0 & \dots & \dots & M_m & \ddots & & \vdots \\ \vdots & \ddots & \ddots & & & \ddots & \ddots & \vdots \\ \vdots & & \ddots & \ddots & & & \ddots & 0 \\ 0 & \dots & \dots & 0 & M_0 & \dots & \dots & M_m \end{bmatrix},$$

where M_i is a $n_r \times n_c$ block-matrix. M is also a *block-matrix* of dimension n(blocks) $\times$ p(blocks).

The block–transposition of a $n.n_r \times p.n_c$ matrix M of block–matrices is denoted by M^b. If M is a $n \times p$ block-matrix of block-matrices M_{ij}, each of size $n_r \times n_c$, then the block–transpose, M^b, consists of $p \times n$ block-matrices, $M^b_{ij} = M_{ji}$, each of size $n_r \times n_c$.

Example 2.3 Consider a 3×2 polynomial matrix

$$L(q)=\begin{bmatrix} q^2+2q+1 & q+4 \\ 2q^2+q+3 & 2q^2+2q \\ q-5 & 3q+1 \end{bmatrix},$$

the row block–matrix description of which is

$$L=\begin{bmatrix} L_0 & L_1 & L_2 \end{bmatrix}=\left[\begin{array}{cc|cc|cc} 1 & 0 & 2 & 1 & 1 & 4 \\ 2 & 2 & 1 & 2 & 3 & 0 \\ 0 & 0 & 1 & 3 & -5 & 1 \end{array}\right].$$

Performing the block–transposition on the row block–matrix L, the column block–matrix description is

$$L^b = \begin{bmatrix} L_0 \\ L_1 \\ L_2 \end{bmatrix} = \begin{bmatrix} 1 & 0 \\ 2 & 2 \\ 0 & 0 \\ - & - \\ 2 & 1 \\ 1 & 2 \\ 1 & 3 \\ - & - \\ 1 & 4 \\ 3 & 0 \\ -5 & 1 \end{bmatrix}.$$

Remark 2.4 The degree of the column or row block–matrix description of a polynomial matrix is invariant under block–transposition, i.e., $\partial[L^b] = \partial[L] = \partial[L(q)]$.

2.4.4 Resultant of a Polynomial Matrix

The multiplication of block–matrices requires the construction of a matrix known as the *resultant* of a polynomial matrix [46,7]. The properties of the resultant of a polynomial matrix will now be established; its use in the multiplication of polynomial matrices will be examined in § 2.4.5 .

Consider a $n_r \times n_c$ polynomial matrix $L(q)$ of degree $\partial[L(q)] = m$, and let $L = [L_0\, L_1\, \ldots\, L_m]$ be the numerical row block–matrix representation of $L(q)$. The resultant of *order* h of the row block–matrix L is denoted by $< L >_h$ and is defined

$$< L >_h \triangleq \sum_{i=0}^{m} R_i \otimes L_i,$$

where:

$\otimes$ denotes the Kronecker product, and
$R_i \triangleq [0_{(h+1)\times i}\; I_{(h+1)}\; 0_{(h+1)\times(m-i)}]$

is the $(h+1) \times (h+m+1)$ matrix containing ones along the $(i+1)$th diagonal and zeros in all other locations ($0_{j\times k}$ denotes a $j \times k$ null matrix). The resultant $< L >_h$ of order

h is a parallelogram block–matrix of polynomial coefficients of the form

$$< L >_h = \begin{bmatrix} L_0 & \dots & \dots & L_m & 0 & \dots & \dots & 0 \\ 0 & L_0 & \dots & \dots & L_m & \ddots & & \vdots \\ \vdots & \ddots & \ddots & & & \ddots & \ddots & \vdots \\ \vdots & & \ddots & \ddots & & & \ddots & 0 \\ 0 & \dots & \dots & 0 & L_0 & \dots & \dots & L_m \end{bmatrix} . \tag{2.5}$$

The resultant $< L >_h$ is a $(h+1) \times (h+m+1)$ matrix of block–matrices, the dimensions of each being $n_r \times n_c$.

The block–transpose of the resultant can be expressed as

$$< L >_h^b \triangleq \sum_{i=0}^{m} R_i^T \otimes L_i,$$

and has the following form

$$< L >_h^b = \begin{bmatrix} L_0 & 0 & \dots & \dots & 0 \\ \vdots & L_0 & \ddots & & \vdots \\ \vdots & \vdots & \ddots & \ddots & \vdots \\ L_m & \vdots & & \ddots & 0 \\ 0 & L_m & & & L_0 \\ \vdots & \ddots & \ddots & & \vdots \\ \vdots & & \ddots & \ddots & \vdots \\ 0 & \dots & \dots & 0 & L_m \end{bmatrix} . \tag{2.6}$$

The following theorem establishes an important property of the resultant of a block–matrix.

Theorem 2.2 The following theorem, [17] lemma G-2, states that a polynomial matrix $L(q)$ is non–singular if and only if the resultant $< L >_h$ of order h is of full row rank for all h; i.e.,

$$\text{rank}(< L >_h) = (h+1).n_r\,.$$

It is apparent that $L(q)$ is non–singular if either L_0 or L_m is non–singular, or equivalently, $L(q)$ is regular or co–regular.

2.4.5 Multiplication of Polynomial Matrices

Theorem 2.3 [45]: Let $C(q)$ and $L(q)$ be two conformable polynomial matrices of the form described by eqn. (2.1). Let

$$C = \left[\begin{array}{cccc} C_0 & C_1 & \dots & C_{\partial[C]} \end{array} \right]$$

and

$$L = \left[\begin{array}{cccc} L_0 & L_1 & \dots & L_{\partial[L]} \end{array} \right]$$

represent the numerical row block-matrices of $C(q)$ and $L(q)$, respectively. Let

$$E = \left[\begin{array}{cccc} E_0 & E_1 & \dots & E_{\partial[E]} \end{array} \right]$$

represent the row block-matrix of coefficients of the product

$$E(q) = C(q).L(q)\,. \tag{2.7}$$

The numerical equivalent of $E(q)$ is realised by constructing the resultant of L of order $\partial[C]$, and premultiplying it by the row block-matrix C; i.e.,

$$E = C. < L >_{\partial[C]}; \tag{2.8}$$

similarly, for the block-transpose,

$$E^b =< C >^b_{\partial[L]} .L^b, \tag{2.9}$$

where

$$\partial[E] \leq \partial[C] + \partial[L]\,.$$

The proof of Theorem 2.3 can be found in Appendix A.

Remark 2.5 It should be noted that the dimensions of the resultant of a polynomial matrix depend both on the *degree* and on the *dimension* of the polynomial matrix. The dimensions of row block-matrices E and C, and the resultant $< L >_{\partial[C]}$ are as follows. Let us denote the row and column dimensions of a matrix by $n_r(\cdot)$ and $n_c(\cdot)$, respectively.

$n_r(< L >_{\partial[C]}) = (\partial[C] + 1).n_r(L(q))$

$$n_c(< L >_{\partial[C]}) = (\partial[C] + \partial[L] + 1).n_c(L(q))$$

$$n_c(C) = (\partial[C] + 1).n_c(C(q))$$

$$n_r(E) = n_r(C)$$

$$n_c(E) = n_c(< L >_{\partial[C]})$$

The row and column dimensions for the block–transpose equivalent are:

$$n_r(< C >^b_{\partial[L^b]}) = (\partial[L^b] + \partial[C] + 1).n_r(C(q))$$

$$n_c(< C >^b_{\partial[L^b]}) = (\partial[L^b] + 1).n_c(C(q))$$

$$n_r(L^b) = (\partial[L^b] + 1).n_r(L(q))$$

$$n_r(E^b) = n_r(< C >^b_{\partial[L^b]})$$

$$n_c(E^b) = n_c(L^b)$$

To illustrate the methodology, let us consider the following example.

Example 2.4 Consider a 2×2 polynomial matrix

$$C(q) = \begin{bmatrix} 3q+2 & 8q^2+6q+1 \\ 4q^3+2q & 8 \end{bmatrix}$$

and a 2×3 polynomial matrix

$$L(q) = \begin{bmatrix} 4q^2+6q+1 & 3q^2+5 & 6q \\ 2q+6 & 9q+2 & 5q^2+2q+1 \end{bmatrix}.$$

The product of the polynomial matrices $C(q)$ and $L(q)$ is to be determined using the above methodology.

The row block–matrix description of $C(q)$ and $L(q)$ are

$$C = \left[\begin{array}{cc|cc|cc|cc} 0 & 0 & 0 & 8 & 3 & 6 & 2 & 1 \\ 4 & 0 & 0 & 0 & 2 & 0 & 0 & 8 \end{array}\right]$$

and

$$L = \left[\begin{array}{ccc|ccc|ccc} 4 & 3 & 0 & 6 & 0 & 6 & 1 & 5 & 0 \\ 0 & 0 & 5 & 2 & 9 & 2 & 6 & 2 & 1 \end{array}\right],$$

respectively.

The resultant of the polynomial matrix $L(q)$ of order $\partial[C(q)] = 3$ is constructed first; i.e.,

$$< L >_3 = \begin{bmatrix} 4 & 3 & 0 & 6 & 0 & 6 & 1 & 5 & 0 & 0 & 0 & 0 & 0 & 0 & 0 & 0 & 0 & 0 \\ 0 & 0 & 5 & 2 & 9 & 2 & 6 & 2 & 1 & 0 & 0 & 0 & 0 & 0 & 0 & 0 & 0 & 0 \\ 0 & 0 & 0 & 4 & 3 & 0 & 6 & 0 & 6 & 1 & 5 & 0 & 0 & 0 & 0 & 0 & 0 & 0 \\ 0 & 0 & 0 & 0 & 0 & 5 & 2 & 9 & 2 & 6 & 2 & 1 & 0 & 0 & 0 & 0 & 0 & 0 \\ 0 & 0 & 0 & 0 & 0 & 0 & 4 & 3 & 0 & 6 & 0 & 6 & 1 & 5 & 0 & 0 & 0 & 0 \\ 0 & 0 & 0 & 0 & 0 & 0 & 0 & 0 & 5 & 2 & 9 & 2 & 6 & 2 & 1 & 0 & 0 & 0 \\ 0 & 0 & 0 & 0 & 0 & 0 & 0 & 0 & 0 & 4 & 3 & 0 & 6 & 0 & 6 & 1 & 5 & 0 \\ 0 & 0 & 0 & 0 & 0 & 0 & 0 & 0 & 0 & 0 & 5 & 2 & 9 & 2 & 6 & 2 & 1 \end{bmatrix}$$

The row block–matrix description of the product $C(q).L(q)$ is

$$C. < L >_{\partial[C]} = \left[\begin{array}{cc|cc|cc|cc} 0 & 0 & 0 & 8 & 3 & 6 & 2 & 1 \\ 4 & 0 & 0 & 0 & 2 & 0 & 0 & 8 \end{array}\right] \times$$

$$\begin{bmatrix} 4 & 3 & 0 & 6 & 0 & 6 & 1 & 5 & 0 & 0 & 0 & 0 & 0 & 0 & 0 & 0 & 0 & 0 \\ 0 & 0 & 5 & 2 & 9 & 2 & 6 & 2 & 1 & 0 & 0 & 0 & 0 & 0 & 0 & 0 & 0 & 0 \\ 0 & 0 & 0 & 4 & 3 & 0 & 6 & 0 & 6 & 1 & 5 & 0 & 0 & 0 & 0 & 0 & 0 & 0 \\ 0 & 0 & 0 & 0 & 0 & 5 & 2 & 9 & 2 & 6 & 2 & 1 & 0 & 0 & 0 & 0 & 0 & 0 \\ 0 & 0 & 0 & 0 & 0 & 0 & 4 & 3 & 0 & 6 & 0 & 6 & 1 & 5 & 0 & 0 & 0 & 0 \\ 0 & 0 & 0 & 0 & 0 & 0 & 0 & 0 & 5 & 2 & 9 & 2 & 6 & 2 & 1 & 0 & 0 & 0 \\ 0 & 0 & 0 & 0 & 0 & 0 & 0 & 0 & 0 & 4 & 3 & 0 & 6 & 0 & 6 & 1 & 5 & 0 \\ 0 & 0 & 0 & 0 & 0 & 0 & 0 & 0 & 0 & 0 & 5 & 2 & 9 & 2 & 6 & 2 & 1 \end{bmatrix}$$

$$= \left[\begin{array}{ccc|ccc|ccc|ccc|ccc|} 0 & 0 & 0 & 0 & 0 & 40 & 28 & 81 & 46 & 86 & 76 & 43 & 53 & 36 & 20 \\ 16 & 12 & 0 & 24 & 0 & 24 & 12 & 26 & 0 & 12 & 0 & 52 & 18 & 82 & 16 \end{array}\right. \cdots$$

$$\cdots \left.\begin{array}{ccc} 8 & 12 & 1 \\ 48 & 16 & 8 \end{array}\right]$$

or, equivalently,

$$C(q).L(q) = \left[\begin{array}{cc} 28q^3 + 86q^2 + 53q + 8 & 81q^3 + 76q^2 + 36q + 12 \\ 16q^5 + 24q^4 + 12q^3 + 12q^2 + 18q + 48 & 12q^5 + 26q^3 + 82q + 16 \end{array}\right. \cdots$$

$$\cdots \left.\begin{array}{c} 40q^4 + 46q^3 + 43q^2 + 20q + 1 \\ 24q^4 + 52q^2 + 16q + 8 \end{array}\right].$$

2.4.6 Synthesis of a Polynomial Block–Matrix Description of a Square Polynomial Matrix with Specified Properties

In devising algorithms for carrying out certain calculations, e.g. transforming a state–space model of a system to its matrix–fraction description, it is necessary to *synthesize* a *square* polynomial matrix $L(q)$ which has specified properties. These properties are given by Definitions 2.10 to 2.11 below.

Proposition 2.3 Let the polynomial matrix $L(q) \in \Re[q]^{n\times n}$ of degree $m = \partial[L(q)]$ and order ν, the jth column index of which is ν_j, $j \in \underline{n}$, be a minimal basis for an n-dimensional column vector space of n-tuples over $\mathcal{F}[q]$.

Definition 2.10 Let the polynomial matrix $L(q)$ be separated into two $n \times n$ polynomial matrices $P_c(q)$ and $E_c(q)$, such that

$$L(q) = P_c(q) + E_c(q), \text{ subject to } \partial_{cj}[P_c(q)] > \partial_{cj}[E_c(q)],$$

where:

$P_c(q)$ will be known as the *column structure* matrix of $L(q)$; and

$E_c(q)$ will be known as the *column residue* matrix of $L(q)$.

Definition 2.11 The column structure matrix, $P_c(q)$, and the column residue matrix, $E_c(q)$, possess the following properties:

1. The column structure matrix, $P_c(q)$, is represented by

$$P_c(q) = \Gamma_c[L(q)].\text{diag}\left[\ q^{\nu_1},\ q^{\nu_2},\ \ldots,\ q^{\nu_n}\ \right], \tag{2.10}$$

   where $\text{diag}[q^{\nu_1}, \ldots, q^{\nu_n}]$ is the $n \times n$ diagonal matrix with diagonal entries q^{ν_j}, $j \in \underline{n}$.

2. Using Definition 2.1 of a polynomial matrix, the general description of the column structure matrix is

$$P_c(q) = \begin{bmatrix} l_{0_{11}}q^{\nu_1} & l_{0_{12}}q^{\nu_2} & \ldots & \ldots & l_{0_{1n}}q^{\nu_n} \\ l_{0_{21}}q^{\nu_1} & l_{0_{22}}q^{\nu_2} & \ldots & \ldots & l_{0_{2n}}q^{\nu_n} \\ \vdots & \vdots & & & \vdots \\ l_{0_{i1}}q^{\nu_1} & l_{0_{i2}}q^{\nu_2} & \ldots & \ldots & l_{0_{in}}q^{\nu_n} \\ \vdots & \vdots & & & \vdots \\ l_{0_{n1}}q^{\nu_1} & l_{0_{n2}}q^{\nu_2} & \ldots & \ldots & l_{0_{nn}}q^{\nu_n} \end{bmatrix}.$$

3. The dimension of $P_c(q)$ is that of $L(q)$; i.e., $n \times n$.

4. $\partial_{cj}[P_c(q)] = \partial_{cj}[L(q)]$, $j \in \underline{n}$.

5. The column residue matrix $E_c(q)$ contains the remaining components of the polynomial elements of $L(q)$ *after* $P_c(q)$ has been extracted from $L(q)$. Thus, the column residue matrix can be described by

$$E_{ij}(q) = \left[l_{1_{ij}} \dots l_{(\nu_j)_{ij}}\right] . Q_{\nu_j - 1}$$

i, $j \in \underline{n}$.

6. Let

$$l(q) = l_0 q^{\nu} + \text{lower degree terms in } q$$

denote the determinant of $L(q)$, the polynomial order of which is ν. From eqn. (2.10) it follows that the determinant of the column structure matrix $P_c(q)$ is $\det(\Gamma_c[L(q)])q^{\nu}$ where

$$\nu = \sum_{j=1}^{n_c} \nu_j .$$

Moreover, the determinant is equal to the highest degree term of $l(q)$, i.e.,

$$\det(P_c(q)) = l_0 q^{\nu},$$

where

$$l_0 = \det(\Gamma_c[L(q)]) .$$

Result 2.3 The highest column-degree coefficient matrix $\Gamma_c[L(q)]$ is non-singular and $\nu = \sum_{j=1}^{n} \nu_j$. Moreover, $L(q)$, which is to be synthesized, is column-proper.

Remark 2.6 It is clear that similar definitions for the structure and residue matrices can be realised when considering the row vector space rather than the column vector space. The terms "$P_c(q)$" and "$E_c(q)$" should be replaced by "$P_r(q)$" and "$E_r(q)$", respectively, when considering the row vector space.

Let us illustrate the above concepts by an example.

Example 2.5 Consider the 3×3 polynomial matrix

$$L(q) = \begin{bmatrix} 5q+1 & q^2+3q+2 & 4q+5 \\ 3q+4 & 2q+1 & q^3+q^2+2 \\ q+7 & 3 & 3 \end{bmatrix} .$$

The column indices are $\nu_1 = 1$, $\nu_2 = 2$, $\nu_3 = 3$ and highest column–degree coefficient matrix is

$$\Gamma_c[L(q)] = \begin{bmatrix} 5 & 1 & 0 \\ 3 & 0 & 1 \\ 1 & 0 & 0 \end{bmatrix} .$$

Thus the column structure matrix is

$$P_c(q) = \begin{bmatrix} 5 & 1 & 0 \\ 3 & 0 & 1 \\ 1 & 0 & 0 \end{bmatrix} .\text{diag}\left[\begin{array}{ccc} q^1, & q^2, & q^3 \end{array}\right]$$

and the column residue matrix is

$$E_c(q) = \begin{bmatrix} 1 & 3q+2 & 4q+5 \\ 4 & 2q+1 & q^2+2 \\ 7 & 3 & 3 \end{bmatrix} .$$

Note that $\Gamma_c[L(q)]$ is non–singular. Therefore $L(q)$ is column proper. Furthermore, the order of $L(q)$ is $\nu = \sum_{j=1}^{3} \nu_j = 6$ which is equal to the polynomial order of

$$\det(L(q)) = q^6 + 11q^5 + 18q^4 + 4q^3 - 12q^2 - 27q + 26 .$$

2.4.6.1 The Skeleton Block–Matrix of a Polynomial Matrix

Proposition 2.4 Consider an unknown $n \times n$ non–singular polynomial matrix $L(q)$ of order ν and degree m, for which the column structure matrix, $P_c(q)$, is specified and the column residue matrix, $E_c(q)$, is unspecified.

The information contained by the column structure matrix, $P_c(q)$, is sufficient to construct the invariant structure of the row block–matrix description of $L(q)$, that is, $P_c(q)$ contains information that defines the basic properties of the polynomial matrix $L(q)$, as described in § 2.4.6. The information contained in the column structure matrix is used to construct a row block–matrix description, Ls, which is known as the *skeleton* row block–matrix of $L(q)$.

Definition 2.12 The skeleton row block–matrix, Ls, is defined as having the following properties:

1. The degree of Ls is that of the column structure matrix $P_c(q)$; i.e., $\partial[Ls] = \partial[P_c(q)] = m$. Thus the skeleton row block–matrix, Ls, is written

$$Ls = \begin{bmatrix} Ls_0 & Ls_1 & \dots & Ls_m \end{bmatrix} .$$

2. The dimension of the skeleton row block–matrix Ls is $n \times n.(m+1)$.

3. The columns of Ls can be divided into two categories, that is, columns associated with the column structure and column residue matrices $P_c(q)$ and $E_c(q)$, respectively.

Let us consider in detail the properties of matrices $P_c(q)$ and $E_c(q)$ and their role in the construction of Ls.

(a) **Columns associated with the column structure matrix $P_c(q)$**

The elements of certain columns of Ls are the coefficients of the polynomial elements in the column structure matrix $P_c(q)$. These columns of Ls can be further divided into the following two subcategories:

1. **Active structure columns**

   There exist n columns of Ls, the elements of which are assigned directly by the highest column–degree coefficient matrix, $\Gamma_c[L(q)]$, of the column structure matrix $P_c(q)$. These columns are known as *active structure columns* of Ls. The active structure columns of Ls are

$$\left\{ \; Ls_{(m-\nu_1)}(:,1), \quad Ls_{(m-\nu_2)}(:,2), \quad \dots, \quad Ls_{(m-\nu_n)}(:,n) \; \right\},$$

   where $Ls_{(m-\nu_j)}(:,j)$ denotes the jth column of the coefficient block–matrix $Ls_{(m-\nu_j)}$ of Ls. Let us consider the jth column $Ls_{(m-\nu_j)}(:,j)$ from the above set of n columns of Ls. The elements of this column directly correspond to the coefficients of the jth column of the highest column–degree coefficient matrix $\Gamma_c[L(q)]$; i.e.,

$$Ls_{(m-\nu_j)}(:,j) = \begin{bmatrix} l_{0_{1j}} \\ \vdots \\ l_{0_{ij}} \\ \vdots \\ l_{0_{nj}} \end{bmatrix},$$

$j \in \underline{n}$. Note that the n active structure columns are linearly independent because $\Gamma_c[L(q)]$ is non–singular.

2. **Inactive structure columns**

Let us consider the jth column polynomial elements of $L(q)$. With reference to Proposition 2.3, the degree of the jth column elements of the n-tuple of $L(q)$ is less than or equal to its jth index ν_j. Therefore, only the jth columns of the coefficient block–matrices $Ls_{(m-\nu_j)}$ to Ls_m of Ls are required to describe the jth column polynomial elements of $L(q)$, when $\nu_j < m$. The jth columns of the remaining block–matrices Ls_0 to $Ls_{(m-\nu_j-1)}$ of Ls contain elements each of which are zero, when $\nu_j < m$. These columns are known as *inactive structure columns* of Ls. For example, if the column structure matrix of $L(q)$, with $m = 2$, is

$$P_c(q) = \begin{bmatrix} 2 & 1 & 0 \\ 3 & 2 & -1 \\ 1 & 0 & 5 \end{bmatrix} .\text{diag}\left[\ q^2,\ q^0,\ q^1\ \right],$$

the indices of which are $\nu_1 = 2$, $\nu_2 = 0$, $\nu_3 = 1$. The highest column–degree coefficient matrix

$$\Gamma_c[L(q)] = \begin{bmatrix} 2 & 1 & 0 \\ 3 & 2 & -1 \\ 1 & 0 & 5 \end{bmatrix}$$

is non–singular. Then

$$Ls = \left[\begin{array}{ccc|ccc|ccc} 2 & 0 & 0 & e & 0 & 0 & e & 1 & e \\ 3 & 0 & 0 & e & 0 & -1 & e & 2 & e \\ 1 & 0 & 0 & e & 0 & 5 & e & 0 & e \end{array}\right].$$

Columns $\{1, 6, 8\}$ of Ls are associated with the active structure columns. The inactive structure columns $\{2, 5\}$ in Ls are associated with the zero coefficients in the ith row polynomials $P_{c_{i2}}(q) = 0.q^2 + 0.q$; likewise the inactive structure column $\{3\}$ is associated with the zero coefficient $P_{c_{i3}}(q) = 0.q^2$, $i \in \underline{n}$.

In forming the skeleton row block–matrix Ls the number of inactive structure columns required by the jth column polynomial elements of $L(q)$ is $t_j = m - \nu_j$. In other words, t_j inactive structure columns are required to blank the jth columns of each of the block–matrices Ls_0 to $Ls_{(t_j-1)}$. Considering all the columns of $L(q)$ we find that the total number of inactive structure columns in the skeleton row block–matrix Ls is:

$$\begin{aligned} t &= \sum_{j=1}^{n} t_j \\ &= (m - \nu_1) + (m - \nu_2) + \cdots + (m - \nu_n) \\ &= n.m - \overbrace{(\nu_1 + \nu_2 + \cdots + \nu_n)}^{\nu} . \end{aligned}$$

However, the sum of the indices ν_j of the n columns of the column structure matrix is equal to the order ν of $L(q)$. Hence

$$t = n.m - \nu .$$

Remark 2.7 In Ls the number of inactive structure columns, t, is the amount by which the order, ν, of $L(q)$ falls short of the maximum possible order, $n.m$, for a polynomial matrix, $L(q)$, with degree m and dimension $n \times n$.

(b) Columns associated with the column residue matrix, $E_c(q)$

These columns of the skeleton block–matrix, which have not been discussed yet, are occupied by the letter 'e' in the example above.

The number of columns of Ls, the elements of which are assigned by the column residue matrix of $L(q)$, can be determined as follows. From property 2 of Definition 2.12, the column dimension of Ls is

$$n_c(Ls) = n.(m+1) .$$

Of these columns, n active columns are assigned to the skeleton row block–matrix Ls by the column structure matrix, $P_c(q)$. Also there are

$$t = n.m - \nu$$

inactive structure columns assigned to Ls. Hence, the number of columns remaining for the assignment by the column residue matrix are

$$n_c(Ls) - n - t = n.m + n - n - (n.m - \nu) = \nu\,.$$

The columns of the skeleton row block–matrix that are assigned by the column residue matrix, $E_c(q)$, are known as the *residue* columns.

Result 2.4 If the polynomial matrix, $L(q)$, is regular then the leading coefficient block–matrix, Ls_0, of its skeleton row block–matrix has a full–rank. Thus eqn. (2.2) satisfies the equality $n.m = \nu$ and $t = 0$; i.e., there are no inactive structure columns in Ls. If $L(q)$ is not regular, then eqn. (2.2) satisfies the inequality, and Ls contains $t = n.m - \nu$ inactive structure columns. Furthermore, the leading coefficient block–matrix Ls_0 is singular.

(c) Examples

The following two examples illustrate the construction and form of the skeleton block–matrix of a polynomial matrix $L(q)$. The first example constructs a skeleton row block–matrix given the column structure matrix, $P_c(q)$, while the second example constructs a skeleton column block–matrix given the row structure matrix $P_r(q)$.

Example 2.6 Let

$$P_c(q) = \begin{bmatrix} 3 & 4 & -1 \\ 1 & 0 & 3 \\ 2 & 5 & -4 \end{bmatrix} .\mathrm{diag}\left[\; q^2, \;\; q^2, \;\; q^1 \;\right]$$

represent the column structure matrix of $L(q)$ of degree $m = 2$ and dimension $n = 3$. The indices of $L(q)$ are $\nu_1 = 2$, $\nu_2 = 2$, $\nu_3 = 1$; the highest column–degree coefficient matrix

$$\Gamma_c[L(q)] = \begin{bmatrix} 3 & 4 & -1 \\ 1 & 0 & 3 \\ 2 & 5 & -4 \end{bmatrix}$$

is non–singular.

There exist $\nu = 5$ residue columns, $n = 3$ active structure columns and $t = n.m - \nu = 3 \times 2 - 5 = 1$ inactive structure column in the skeleton row block–matrix description of $L(q)$. The skeleton row block–matrix of $L(q)$ can now be constructed as follows:

$$Ls = \begin{bmatrix} 3 & 4 & 0 & | & e & e & -1 & | & e & e & e \\ 1 & 0 & 0 & | & e & e & 3 & | & e & e & e \\ 2 & 5 & 0 & | & e & e & -4 & | & e & e & e \end{bmatrix}.$$

The $n = 3$ active structure columns and $t = 1$ inactive structure column of Ls, assigned by the column structure matrix $P_c(q)$, are $\{1, 2, 6\}$ and $\{3\}$, respectively. The $\nu = 5$ residue columns are $\{4, 5, 7, 8, 9\}$; these columns are denoted by the letter 'e' in the skeleton row block–matrix. Hence, $L(q)$ can be written

$$L(q) = \begin{bmatrix} 3q^2 + E_{c_{11}}(q) & 4q^2 + E_{c_{12}}(q) & -q + E_{c_{13}}(q) \\ q^2 + E_{c_{21}}(q) & E_{c_{22}}(q) & 3q + E_{c_{23}}(q) \\ 2q^2 + E_{c_{31}}(q) & 5q^2 + E_{c_{32}}(q) & -4q + E_{c_{33}}(q) \end{bmatrix},$$

where $E_{c_{ij}}(q)$ denotes the ith row, jth column, polynomial element of the column residue matrix.

The degree of the polynomial elements of columns 1, 2 and 3 of the column residue matrix are 1 or less, 1 or less and 0, respectively.

Example 2.7 Let

$$P_r(q) = \text{diag}\left[\ q^2,\ \ q^3,\ \ q^1\ \right] \cdot \begin{bmatrix} 8 & 0 & 0 \\ 0 & 2 & 0 \\ 0 & 0 & 1 \end{bmatrix}$$

represent the row structure matrix of $L(q)$ of degree $m = 3$ and dimension $n = 3$. The indices of $L(q)$ are $\mu_1 = 2$, $\mu_2 = 3$, $\mu_3 = 1$; the highest row–degree coefficient matrix

$$\Gamma_r[L(q)] = \begin{bmatrix} 8 & 0 & 0 \\ 0 & 2 & 0 \\ 0 & 0 & 1 \end{bmatrix}$$

is full–rank.

There exist $\mu = 6$ residue rows, $n = 3$ active structure rows and $t = n.m - \mu = 3 \times 3 - 6 = 3$ inactive structure rows in the skeleton column block–matrix of $L(q)$. The skeleton column block–matrix of $L(q)$ is thus:

$$L^b s = \begin{bmatrix} 0 & 0 & 0 \\ 0 & 2 & 0 \\ 0 & 0 & 0 \\ - & - & - \\ 8 & 0 & 0 \\ e & e & e \\ 0 & 0 & 0 \\ - & - & - \\ e & e & e \\ e & e & e \\ 0 & 0 & 1 \\ - & - & - \\ e & e & e \\ e & e & e \\ e & e & e \end{bmatrix}.$$

The $n = 3$ active structure rows and $t = 3$ inactive structure rows of $L^b s$, assigned by the row structure matrix $P_r(q)$, are $\{2, 4, 9\}$ and $\{1, 3, 6\}$, respectively. The $\nu = 6$ residue rows are $\{5, 7, 8, 10, 11, 12\}$. The position of the $t = 3$ inactive structure rows ensure that the polynomial elements of rows 1 and 3 do not exceed a degree of 2 and 1, respectively. Hence, $L(q)$ can be written

$$L(q) = \begin{bmatrix} 8q^2 + E_{11}(q) & E_{12}(q) & E_{13}(q) \\ E_{21}(q) & 2q^3 + E_{22}(q) & E_{23}(q) \\ E_{31}(q) & E_{32}(q) & q + E_{33}(q) \end{bmatrix}.$$

The degrees of the polynomial elements of rows 1, 2 and 3 of the row residue matrix are 1 or less, 2 or less and 0, respectively.

Remark 2.8 It can be seen from these examples that a column structure matrix, $P_c(q)$, generates a skeleton row block–matrix, the polynomial matrix of which is column–proper. Likewise, a row structure matrix, $P_r(q)$, generates a skeleton column block–matrix, the polynomial matrix of which is row–proper.

Remark 2.9 It is clear that the properties of skeleton row block–matrix, Ls, which are defined for row block–matrix descriptions of the polynomial matrix, $L(q)$, equally apply to the skeleton column block–matrix description in which the row structure and row residue matrices are employed.

2.4.6.2 A 'Standard' Structure Matrix and its Skeleton Block–Matrix

Let us consider a $n \times n$ column–proper polynomial matrix $L(q)$ of order ν, the column structure matrix of which is $P_c(q)$. An infinite number of column structure matrices, and thus an infinite number of skeleton row block–matrices, can be found that satisfy the constraints of dimension, n, and order, ν, placed on the polynomial matrix $L(q)$. However, we propose to define a form for both the column structure matrix and its skeleton row block–matrix; we will call this the 'standard form' of these two matrices. Let us consider the following proposition.

Proposition 2.5 It is proposed that (i) the standard form of the column structure matrix and its skeleton row block–matrix be of minimal degree m, (ii) the skeleton row block–matrix possess a minimal number of inactive structure columns (i.e., $t = n.m - \nu$ is minimal), (iii) the indices ν_j are, as closely as possible, the same value, and (iv) the active structure columns consist of unitary columns, i.e., the highest column–degree coefficient matrix, $\Gamma_c[L(q)]$, is a unitary matrix.

It is known that the degree, m, of a polynomial matrix $L(q)$ is related to its dimension, n, and polynomial order, ν, by the following inequality (see Result 2.2):

$$n.m \geq \nu \,.$$

The minimal degree a $n \times n$ polynomial matrix $L(q)$, of polynomial order ν, is the degree m which satisfies the above inequality. Because the number of inactive structure columns contained by a skeleton row block–matrix is

$$t = n.m - \nu,$$

then, by choosing the minimal degree m for $L(q)$, the number of inactive structure columns, t, of the skeleton row block–matrix is also minimised.

Consider a *regular* column structure matrix of order $n.m$ and degree m, the n indices of which are $\nu_1 = \cdots = \nu_n = m$; further let $\nu_{\text{reg}} = \sum_{i=1}^{n} \nu_i = n.m$. Thus

$$t = \nu_{\text{reg}} - \nu, \tag{2.11}$$

where $0 \leq t \leq (n-1)$. Equation (2.11) denotes the amount by which a column structure matrix, $P_c(q)$, of order ν falls short of the order, $\nu_{\text{reg}} = n.m$, of a regular column structure matrix of dimension n and degree m. To transform a regular column structure matrix of order ν_{reg} into a column structure matrix of order ν, we have to, according to eqn. (2.11), reduce the order by $t = \nu_{\text{reg}} - \nu$ while maintaining the minimal degree m. This can be accomplished by reducing, say, the first t indices of the regular column structure matrix to a degree of $m-1$ with the remaining $n-t$ indices unaltered. Thus the order of the regular structure matrix is reduced to a column structure matrix of order $n.m - t = \nu$. Consequently we have a column structure matrix of minimal degree m, the indices of which are $\nu_1 = \cdots = \nu_t = (m-1)$ and $\nu_{(t+1)} = \cdots = \nu_n = m$, where $\sum_{j=1}^{n} \nu_j = (m-1).t + (n-t).m = \nu$. Therefore, a $n \times n$ column structure matrix of order ν and minimal degree m can be realised using indices the first t and remaining $(n-t)$ values of which are $(m-1)$ and m, respectively.

In order to provide a simple structure for Ls, the highest column–degree coefficient matrix is chosen to be

$$\Gamma_c[L(q)] = \begin{bmatrix} 0 & I_{(n-t)} \\ I_t & 0 \end{bmatrix}$$

where I_t is the $t \times t$ identity matrix.

We should also note that

$$\det(L(q)) = q^{\nu} + \text{lower degree terms in } q$$

is monic; i.e., $\det(\Gamma_c[L(q)]) = 1$.

The standard form of the column structure matrix is thus

$$P_c(q) = \begin{bmatrix} 0 & q^{m} I_{(n-t)} \\ q^{(m-1)} I_t & 0 \end{bmatrix}.$$

Algorithm 2.1 The following steps summarise the methodology used to construct a standard form of the column structure matrix and its skeleton row block–matrix:

step 1. Compute the minimal degree, m, of the column–proper polynomial matrix $L(q)$, the dimension and order of which are n and ν, respectively, by minimising the inequality

$$n.m \geq \nu .$$

step 2. Set $t = n.m - \nu$.

step 3. Construct the standard form of the column structure matrix

$$P_c(q) = \begin{bmatrix} 0 & q^m I_{(n-t)} \\ q^{(m-1)} I_t & 0 \end{bmatrix},$$

the skeleton row block–matrix of which is

$$Ls = \begin{bmatrix} 0_t & I_n & E_\nu \end{bmatrix},$$

where:

0_t is the t-column matrix of zeros, the columns of which represent the t inactive structure columns of Ls;

I_n is the $n \times n$ identity matrix, the columns of which represent the n active structure columns of Ls; and

E_ν is the ν-column matrix, the columns of which represents the ν residue columns of Ls.

To illustrate the standard form and construction of the column structure matrix and its skeleton row block–matrix, let us consider the following example.

Example 2.8 Let $L(q)$ be a 3×3 polynomial matrix of polynomial order $\nu = 7$.

step 1. Compute the minimal degree m of the column structure matrix, $P_c(q)$, using the inequality

$$(n = 3).m \geq (\nu = 7).$$

Hence $m = 3$.

step 2. The number of inactive structure columns, t, contained in the skeleton row block–matrix is

$$t = n.m - \nu = 3 \times 3 - 7 = 2.$$

step 3. Construct the standard form of the column structure matrix; i.e.,

$$P_c(q) = \begin{bmatrix} 0 & q^m I_{(n-t)} \\ q^{(m-1)} I_t & 0 \end{bmatrix}$$

$$= \begin{bmatrix} 0 & 0 & q^3 \\ q^2 & 0 & 0 \\ 0 & q^2 & 0 \end{bmatrix}.$$

Thus, the skeleton row block–matrix is

$$Ls = \left[\begin{array}{ccc|ccc|ccc|ccc} 0 & 0 & 1 & 0 & 0 & e & e & e & e & e & e & e \\ 0 & 0 & 0 & 1 & 0 & e & e & e & e & e & e & e \\ 0 & 0 & 0 & 0 & 1 & e & e & e & e & e & e & e \end{array}\right].$$

Remark 2.10 The properties of the standard form of the column structure matrix and its skeleton row block–matrix can be equally applied to the row structure matrix and its skeleton column block–matrix.

2.4.7 Construction of a Non–Singular Polynomial Matrix with a Specified Set of Eigenvalues

In the design process, described in Chapter 4., it is necessary to construct a non–singular polynomial matrix with a specified set of eigenvalues. In this section, we propose to develop a methodology that enables us to construct such a polynomial matrix.

More specifically, we need to construct a non–singular polynomial matrix, the column structure matrix of which is in standard form. Therefore, we shall restrict ourselves to considering a non–singular polynomial matrix $L(q)$, the column structure matrix of which is

$$P_c(q) = \left(\Gamma_c[L(q)] = \begin{bmatrix} 0 & I_{(n-t)} \\ I_t & 0 \end{bmatrix}\right).\text{diag}\begin{bmatrix} q^{\nu_1}, & q^{\nu_2}, & \ldots, & q^{\nu_n} \end{bmatrix}.$$

The column indices of $L(q)$ are $\nu_1 = \cdots = \nu_t = (m-1)$ and $\nu_{t+1} = \cdots = \nu_n = m$, where $t = n.m - \nu$ and $0 \leq t \leq (n-1)$. The order of $L(q)$ is $\nu = \sum_{j=1}^{n} \nu_j$ (see § 2.4.6.2).

Comment: A polynomial matrix, $L(q)$, constrained to possess a specified set of eigenvalues and a specified column structure matrix is not unique. However, using the standard form of the column structure matrix a non–singular polynomial matrix of least degree is constructed.

∇∇∇

Let us further constrain the polynomial matrix L(q) to be of the form

$$L(q) = \Gamma_c[L(q)].\text{diag}\left[\; q^{\nu_1} + E_1(q),\;\; q^{\nu_2} + E_2(q),\;\; \ldots,\;\; q^{\nu_n} + E_n(q) \;\right] \tag{2.12}$$

where $E_j(q)$, $j \in \underline{n}$, denotes the non–zero residue polynomial elements of $L(q)$ and $\det(L(q)) = (\det(\Gamma_c[L(q)]) = 1).\prod_{j=1}^{n}(q^{\nu_j} + E_j(q))$.

Let us initially consider the following three steps which describe a simple methodology that can be used to realise a polynomial of the form shown in eqn. (2.12).

step 1. Systematically assign ν_j, $j \in \underline{n}$, eigenvalues to the jth diagonal polynomial element $q^{\nu_j} + E_j(q)$ of $L(q)$. Complex conjugate pairs must be assigned in pairs to realise a real polynomial matrix.

step 2. Compute the n diagonal polynomial elements $q^{\nu_j} + E_j(q)$, $j \in \underline{n}$, of $L(q)$, the zeros of which are the eigenvalues assigned in step 1.

step 3. Construct the polynomial matrix

$$L(q) = \Gamma_c[L(q)].\text{diag}\left[\; q^{\nu_1} + E_1(q),\;\; q^{\nu_2} + E_2(q),\;\; \ldots,\;\; q^{\nu_n} + E_n(q) \;\right].$$

Example 2.9 Let us consider a 3×3 polynomial matrix $L(q)$ of order $\nu = 8$, the eigenvalues of which are:

$$-1, \quad -3, \quad -2 \pm 2.5j, \quad 0 \pm 3j \text{ and } \quad 3 \pm 4j.$$

In this case $m = 3$ and $t = 1$, hence the standard form of the column structure matrix is

$$P_c(q) = \begin{bmatrix} 0 & 1 & 0 \\ 0 & 0 & 1 \\ 1 & 0 & 0 \end{bmatrix} .\text{diag}\left[\; q^2 \;\; q^3 \;\; q^3 \;\right],$$

where $\nu_1 = 2$, $\nu_2 = \nu_3 = 3$. We can begin by assigning $\nu_1 = 2$, $\nu_2 = 3$ and $\nu_3 = 3$ eigenvalues to the polynomial elements $q^2 + E_1(q)$, $q^3 + E_2(q)$ and $q^3 + E_3(q)$, respectively. An appropriate assignment of eigenvalues to the diagonal polynomial elements $q^2 + E_1(q)$,

$q^3 + E_2(q)$ and $q^3 + E_3(q)$, such that a real polynomial matrix is realised, is $\{-2 \pm 2.5j\}$, $\{-1,\ 0 \pm 3j\}$ and $\{-3,\ 3 \pm 4j\}$, respectively. Thus, the resulting polynomial matrix is

$$L(q) = \begin{bmatrix} 0 & q^3 + q^2 + 9q + 9 & 0 \\ 0 & 0 & q^3 - 3q^2 + 7q + 75 \\ q^2 + 4q + 10.25 & 0 & 0 \end{bmatrix}.$$

In certain cases this straightforward method which is used in Example 2.9 must be modified. Consider the following example.

Example 2.10 Let us consider a polynomial matrix $L(q)$ of order $\nu = 8$, the set of eigenvalues of which are:

$$0 \pm j, \quad 1 \pm j, \quad -2 \pm 3j \text{ and } \quad 3 \pm 4j.$$

The standard form of the column structure matrix of $L(q)$ is identical to that of Example 2.9. With this combination of column structure matrix and eigenvalues, the method used in Example 2.9 obviously does not allow us to realise a real polynomial matrix. Moreover, the form of the column structure matrix forces one of the complex conjugate pairs to be separated and assigned to two separate polynomial elements, thus generating a complex coefficient polynomial matrix. Thus, to construct a real polynomial matrix the above method must be modified. To overcome this problem it is proposed that the polynomial matrix $L(q)$ assume the following form.

$$L(q) =$$

$$\Gamma_c[L(q)].\text{diag}\left[\ q^2 + \tilde{E}_1(q),\quad q^2 + \tilde{E}_2(q),\quad q^2 + \tilde{E}_3(q)\ \right].\begin{bmatrix} 1 & 0 & 0 \\ 0 & (q-r) & -i \\ 0 & i & (q-r) \end{bmatrix},$$

where r and i denote the real and complex part, respectively, of the yet unassigned complex conjugate pair.

An appropriate assignment of eigenvalues to the polynomial elements $q^2 + \tilde{E}_1(q)$, $q^2 + \tilde{E}_2(q)$ and $q^2 + \tilde{E}_3(q)$, such that a real polynomial matrix is realised, is $\{-2 \pm 3j\}$, $\{0 \pm j\}$ and $\{1 \pm j\}$, respectively. The remaining complex conjugate pair which is unassigned is $\{3 \pm 4j\}$. Thus, the resulting polynomial matrix $L(q)$ is as follows.

$$L(q) = \begin{bmatrix} 0 & 1 & 0 \\ 0 & 0 & 1 \\ 1 & 0 & 0 \end{bmatrix}.\text{diag}\left[\ q^2 + 4q + 13,\quad q^2 + 1,\quad q^2 - 2q + 2\ \right].\begin{bmatrix} 1 & 0 & 0 \\ 0 & q-3 & -4 \\ 0 & 4 & q-3 \end{bmatrix}$$

or, equivalently,

$$L(q) = \begin{bmatrix} 0 & q^3 - 3q^2 + q - 3 & -4q^2 - 4 \\ 0 & 4q^2 - 8q + 8 & q^3 - 5q^2 + 8q - 6 \\ q^2 + 4q + 13 & 0 & 0 \end{bmatrix}.$$

Algorithm 2.2 The following steps summarise a complete methodology, which combines the ideas discussed above to realise a polynomial matrix, the column structure matrix and eigenvalues of which are specified. Further, let $L(q) = \tilde{L}(q).\overline{L}(q)$, where $\tilde{L}(q)$ and $\overline{L}(q)$ are defined in the following steps 1 and 5, respectively.

step 1. Systematically distribute among the n diagonal polynomial elements $q^{\tilde{\nu}_j} + \tilde{E}_j(q)$, $j \in \underline{n}$, of the polynomial matrix

$$\tilde{L}(q) = \Gamma_c[L(q)].\mathrm{diag}\left[\; q^{\tilde{\nu}_1} + \tilde{E}_1(q),\; q^{\tilde{\nu}_2} + \tilde{E}_2(q),\; \ldots,\; q^{\tilde{\nu}_n} + \tilde{E}_n(q) \;\right], \quad (2.13)$$

beginning with the first diagonal polynomial element, the maximum number of complex conjugate pair eigenvalues to each of the polynomial elements such that the resulting polynomial matrix $\tilde{L}(q)$ is real and $\tilde{\nu}_j \leq \nu_j$, $j \in \underline{n}$.

step 2. As in step 1, distribute among the n diagonal polynomial elements $q^{\tilde{\nu}_j} + \tilde{E}_j(q)$, $j \in \underline{n}$, of $\tilde{L}(q)$ as many as possible of the real eigenvalues to each of the polynomial elements such that $\tilde{\nu}_j \leq \nu_j$.

step 3. Compute the n polynomial elements $q^{\tilde{\nu}_j} + \tilde{E}_1(q)$, $j \in \underline{n}$, of $\tilde{L}(q)$, the zeros of which have been assigned in steps 1 and 2.

step 4. Construct the polynomial matrix $\tilde{L}(q)$, the elements of which correspond to the n diagonal polynomial elements computed in step 3.

step 5. If all complex conjugate pairs have been assigned to the n diagonal polynomial elements of $\tilde{L}(q)$ during the execution of step 1, the algorithm terminates and $L(q) =$

$\tilde{L}(q)$. Otherwise construct the $n \times n$ polynomial matrix

$$\overline{L}(q) = \begin{bmatrix} 1 & & & & & & & & \\ & \ddots & & & & & 0 & & \\ & & 1 & & & & & & \\ & & & q-r & & -i & & & \\ & & & & 1 & & & & \\ & & & & & \ddots & & & \\ & & & i & & q-r & & & \\ & & & & & & 1 & & \\ & 0 & & & & & & \ddots & \\ & & & & & & & & 1 \end{bmatrix}, \qquad (2.14)$$

where r and i denote the real and complex part, respectively, of the complex conjugate pair which is yet unassigned. The terms $q - r$ in eqn. (2.14) lie along the main diagonal, positioned in the sth and wth column of $\overline{L}(q)$, where s and w correspond to the two columns of $\tilde{L}(q)$, the degree of the polynomial elements of which are $\tilde{\nu}_s = \nu_s - 1$ and $\tilde{\nu}_w = \nu_w - 1$, respectively. With reference to eqn. (2.14) the imaginary components i and $-i$ are positioned in (w, s) and (s, w), respectively.

step 6. Construct $L(q) = \tilde{L}(q).\overline{L}(q)$.

To illustrate the algorithm, let us consider the following example which makes use of the numerical techniques of polynomial matrices to construct the polynomial matrix $L(q)$, the column structure matrix and eigenvalues of which are specified below. The computations are carried out on with the aid of a software package [53] and double precision arithmetic.

Example 2.11 Let us consider a 3×3 polynomial matrix $L(q)$ of order $\nu = 4$, the column structure matrix of which is

$$P_c(q) = \begin{bmatrix} 0 & 0 & 1 \\ 1 & 0 & 0 \\ 0 & 1 & 0 \end{bmatrix} .\text{diag}\left[\begin{array}{ccc} q, & q, & q^2 \end{array}\right]$$

with $\nu_1 = \nu_2 = 1$ and $\nu_3 = 2$. The eigenvalues of $L(q)$ are:

$$-1 \pm j \text{ and } \quad 2 \pm j\,.$$

step 1. The eigenvalues of $L(q)$ contain two complex conjugate pairs. The 1st and 2nd diagonal polynomial elements $q^{\tilde{\nu}_1}+\tilde{E}_1(q)$ and $q^{\tilde{\nu}_2}+\tilde{E}_2(q)$ of $\tilde{L}(q)$ can have a maximum of one eigenvalue assigned to each of them. Thus, no complex conjugate pairs can be assigned to the 1st and 2nd diagonal polynomial elements. Hence $\tilde{\nu}_1 = \nu_1 - 1 = 0$ and $\tilde{\nu}_2 = \nu_2 - 1 = 0$. The 3rd diagonal polynomial element $q^{\tilde{\nu}_3} + \tilde{E}_3(q)$ of $\tilde{L}(q)$ can have a maximum of two eigenvalues assigned to it. Thus one complex conjugate pair can be assigned to the 3rd diagonal polynomial element, say, $\{-1 \pm j\}$. Hence $\tilde{\nu}_3 = \nu_3 = 2$.

step 2. There are no real eigenvalues.

step 3. Using a routine provided by the software package, the coefficients of the polynomial elements $q^{\tilde{\nu}_j} + \tilde{E}_j(q)$, $j \in \underline{n}$, are either assigned or computed. Thus the three non-zero polynomial elements of $\tilde{L}(q)$ are:

$$\begin{aligned} q^{\tilde{\nu}_1=0} + \tilde{E}_1(q) &= 1 \\ q^{\tilde{\nu}_2=0} + \tilde{E}_2(q) &= 1 \\ q^{\tilde{\nu}_3=2} + \tilde{E}_3(q) &= q^2 + 2q + 2\,. \end{aligned}$$

step 4. The polynomial matrix $\tilde{L}(q)$ is of the form

$$\tilde{L}(q) = \begin{bmatrix} 0 & 0 & 1 \\ 1 & 0 & 0 \\ 0 & 1 & 0 \end{bmatrix} .\mathrm{diag}\left[\ 1,\ \ 1,\ \ q^2+2q+2\ \right],$$

the row block–matrix description of which is

$$\tilde{L} = \left[\begin{array}{ccc|ccc|ccc} 0 & 0 & 1 & 0 & 0 & 2 & 0 & 0 & 2 \\ 0 & 0 & 0 & 0 & 0 & 0 & 1 & 0 & 0 \\ 0 & 0 & 0 & 0 & 0 & 0 & 0 & 1 & 0 \end{array}\right].$$

step 5. The 1st and 2nd column polynomial elements of $\tilde{L}(q)$ are of degree $\tilde{\nu}_1 = \nu_1 - 1 = 0$ and $\tilde{\nu}_2 = \nu_2 - 1 = 0$, thus the sth and wth columns are 1 and 2, respectively, and the yet unassigned complex conjugate pair is $\{2 \pm j\}$. Thus, the polynomial matrix $\overline{L}(q)$ is of the form

$$\overline{L}(q) = \begin{bmatrix} q-2 & -1 & 0 \\ 1 & q-2 & 0 \\ 0 & 0 & 1 \end{bmatrix},$$

the row block–matrix description of which is

$$\overline{L} = \begin{bmatrix} 1 & 0 & 0 & | & -2 & -1 & 0 \\ 0 & 1 & 0 & | & 1 & -2 & 0 \\ 0 & 0 & 0 & | & 0 & 0 & 1 \end{bmatrix}.$$

step 6. Using the method of resultants we can proceed as follows.

$$L = \tilde{L}. < \overline{L} >_{\partial[\tilde{L}]}$$

$$= \begin{bmatrix} 0 & 0 & 1 & | & 0 & 0 & 2 & | & 0 & 0 & 2 \\ 0 & 0 & 0 & | & 1 & 0 & 0 & | & -2 & -1 & 0 \\ 0 & 0 & 0 & | & 0 & 1 & 0 & | & 1 & -2 & 0 \end{bmatrix}$$

or, equivalently,

$$L(q) = \begin{bmatrix} 0 & 0 & q^2 + 2q + 2 \\ q - 2 & -1 & 0 \\ 1 & q - 2 & 0 \end{bmatrix}.$$

2.4.7.1 Summary

In step 1 and step 2 of Algorithm 2.2, there is no preferred way of assigning the eigenvalues, since there is no one unique polynomial matrix which satisfies the constraints of its specified column structure matrix and eigenvalues. The numerical methodology uses commercially–available software routines and the method of resultants.

2.4.8 The Determinant of a Polynomial Matrix

The design procedure, described in Chapter 4, requires an algorithm which enables the designer to compute the determinant polynomial, $l(q)$, of a square polynomial matrix, $L(q)$, given its row block–matrix description, L.

There are a number of methods available for obtaining the determinant of a square polynomial matrix given its numerical representation, for example, (i) by applying the techniques of interpolation, or (ii) by constructing the companion matrix of the polynomial matrix $L(q)$. The first methodology is not numerically reliable since it relies on the Vandermonde matrix which, in some instances, may be ill–conditioned, especially for polynomials matrices when the polynomial order is greater than 4 [20].

The numerical method proposed for realising the determinant of a polynomial matrix, $L(q)$, utilises the standard representation of a companion matrix, the characteristic polynomial of which is the determinant polynomial, $l(q)$, being sought. The main reason for choosing this methodology is that the software package [53] provides a numerically robust routine for computing the characteristic polynomial, $\det(q.I - C_m(L(q)))$, of a square, first degree polynomial matrix $q.I - C_m(L(q))$, where $C_m(L(q))$ denotes the companion matrix of a monic polynomial matrix $L(q)$.

The proposed methodology can only be implemented on polynomial matrices that are regular; i.e., the coefficient block–matrix L_0 of $L(q)$ must be non–singular. If the polynomial matrix $L(q)$ is not regular a recursive algorithm described in Appendix C is applied to transform $L(q)$ to a regular polynomial matrix $L(q).T(q)$. A simple relationship exists between the determinant polynomial $\det(L(q))$ and the determinant polynomial $\det(L(q).T(q))$.

Consider the standard description of the $n \times n$ polynomial matrix (see Definition 2.2)

$$L(q) = \begin{bmatrix} L_0 & L_1 & L_2 & L_3 & \dots & L_m \end{bmatrix} . \left(\begin{bmatrix} q^m \\ q^{m-1} \\ q^{m-2} \\ q^{m-3} \\ \vdots \\ 1 \end{bmatrix} \otimes I_n \right),$$

the rank and degree of which are n and m, respectively, and I_n is the $n \times n$ identity matrix.

The methodology is considered in two parts. Firstly, we consider the case when the polynomial matrix $L(q)$ is regular and thus L_0 non–singular; secondly, we deal with the case when the polynomial matrix is not regular, i.e., L_0 is singular.

Case 1: L_0 non–singular

Assume $L(q)$ is regular. We first form its monic polynomial matrix equivalent

$$\overline{L}(q) = \begin{bmatrix} I_n & \overline{L}_1 & \overline{L}_2 & \overline{L}_3 & \dots & \overline{L}_m \end{bmatrix} . \left(\begin{bmatrix} q^m \\ q^{m-1} \\ q^{m-2} \\ q^{m-3} \\ \vdots \\ 1 \end{bmatrix} \otimes I_n \right),$$

such that $L(q) = L_0.\overline{L}(q)$. It is well–known [46] that the roots of the monic polynomial matrix $\overline{L}(q)$ are the eigenvalues of its $n.m \times n.m$ first degree monic polynomial matrix

$$q.I_{m.n} - C_m(\overline{L}(q))$$

where

$$C_m(\overline{L}(q)) = \begin{bmatrix} -\overline{L}_1 & -\overline{L}_2 & -\overline{L}_3 & \dots & \dots & -\overline{L}_m \\ I_n & 0_n & 0_n & \dots & \dots & 0_n \\ 0_n & I_n & 0_n & & & \vdots \\ 0 & 0 & I_n & & & \vdots \\ \vdots & \vdots & \ddots & \ddots & & \vdots \\ 0_n & 0_n & \dots & 0_n & I_n & 0_n \end{bmatrix}$$

Hence

$$\det(L_0).\det(q.I_{m.n} - C_m(\overline{L}(q)) = \det(L(q)) .$$

The routine $POLY(\cdot)$ [53] computes the vector of coefficients of the characteristic polynomial of a first degree, monic, square polynomial matrix. By application of this routine to the companion matrix $C_m(\overline{L}(q))$ the determinant polynomial is realised. [$POLY(\cdot)$ uses the numerically robust routines in the package EISPACK [109] to compute the characteristic polynomial of $q.I_{m.n} - C_m(\overline{L}(q))$.]

Case 2: L_0 singular

Assume the polynomial matrix $L(q)$ is not regular. A companion matrix of the polynomial matrix $L(q)$ cannot be implemented directly, since the leading coefficient block–matrix, L_0, is singular. The polynomial matrix $L(q)$ can be transformed into a regular polynomial matrix by postmultiplying it by a properly chosen polynomial matrix $T(q)$ such that $L^{(p)}(q) = L(q).T(q)$ is regular [90]. It is proposed that the polynomial matrix $T(q)$

be chosen such that a simple relationship exists between the determinant polynomials of $L(q)$ and $L^{(p)}(q)$. Hence it is easy to extract the determinant polynomial of $L(q)$ from the determinant polynomial of $L^{(p)}(q)$. A polynomial matrix $T(q)$ which satisfies the above two requirements can be realised by applying Algorithm C.1 in Appendix C. The resulting polynomial matrix $T(q)$ evaluated by this algorithm has the following desirable properties: $T(q)$ is nilpotent; i.e., $\det(T(q)) = c.q^k$, $L^{(p)}(q) = L(q).T(q)$ is regular, and $\det(L^{(p)}(q)) = \det(L(q)).\det(T(q)) = l(q).c.q^k$ (c is a constant). Thus, because the determinant polynomials of $L(q)$ and $L^{(p)}(q)$ are related by a very simple expression, the determinant polynomial of $L(q)$ can be realised easily once the determinant polynomial of $L^{(p)}(q)$ has been calculated.

It is suggested, to obtain a clear understanding of the following algorithm, that the reader first examine Appendix C.

Algorithm 2.3 The row block–matrix description, L, of $L(q)$ is given. In order to compute the determinant, $l(q) = \det(L(q))$, of $L(q)$ using the methodology described above, the following steps are proposed:

step 1. Set $i = 1$.

step 2. Set $L^{(0)} = L$, $k_0 = 0$.

step 3. Set $r_i = \text{rank}(L_0^{(i-1)})$.

step 4. If $r_i = n$, then decrease i by 1 and go to step 11.

step 5. Set $k_i = n - r_i$

step 6. Construct the row block matrix description $U^{(i)}$ of the column shift polynomial matrix (c.s.p.m.) of order k_i; i.e., $U^{(i)} = [0_{r_i}\ I_n\ 0_{k_i}]$.

step 7. Using the routine $QR(\cdot)$, which is provided by commercial software packages, perform QR decomposition on $L_0^{(i)}$ so as to determine the non–singular matrix $Q^{(i)}$, such that $L_0^{(i-1)} = [0_{k_i}\ L_0^{(i)}].(Q^{(i)})^{-1}$.

step 8. Set $L^{(i)} = L^{(i-1)}. < Q^{(i)}.U^{(i)} >_{\partial[L^{(i-1)}]}$.

step 9. Increase i by 1.

step 10. Go to step 3.

step 11. Set $\overline{L} = (L_0^{(i)})^{-1}.L^{(i)}$.

step 12. Construct the companion matrix $C_m(\overline{L}(q))$ of the monic polynomial matrix $\overline{L}(q)$; i.e.,

$$C_m(\overline{L}(q)) = \begin{bmatrix} -\overline{L}_1 & -\overline{L}_2 & -\overline{L}_3 & \dots & \dots & -\overline{L}_m \\ I_n & 0_n & 0_n & \dots & \dots & 0_n \\ 0_n & I_n & 0_n & & & \vdots \\ 0_n & 0_n & I_n & & & \vdots \\ \vdots & \vdots & \ddots & \ddots & & \vdots \\ 0_n & 0_n & \dots & 0_n & I_n & 0_n \end{bmatrix}.$$

step 13. Compute the coefficient vector $l^{(i)}$ of the characteristic polynomial of the companion matrix $C_m(\overline{L}(q))$, using the routine $POLY(\cdot)$; i.e., $l^{(i)} = POLY(C_m(\overline{L}(q)))$.

step 14. Set $k = \sum_{j=1}^{i} k_j$, $c = \prod_{j=1}^{i} DET(Q^{(j)})$.

step 15. Set

$$l = \frac{DET(L_0^{(i)})}{c} . \left[\, l_0^{(i)} \quad \dots \quad l_{\partial[l]-k}^{(i)} \, \right],$$

where $DET(\cdot)$ is a software routine [53] which calculates the determinant of a real matrix.

Example 2.12 Consider the polynomial matrix $L(q)$ which is not regular and is of degree $\partial[L(q)] = 2$; its row block–matrix L shown below. The determinant polynomial $l(q)$ is to be determined using the above methodology. The computations are carried out with the aid of a software package using double precision arithmetic.

Let

$$L = L^{(0)} = \left[\begin{array}{ccc|ccc|ccc} 0 & 0 & 4 & 1 & 2 & 11.96 & 71.03 & 57.38 & -24.65 \\ 0 & 0 & 2 & 1 & 1 & 4.62 & 32.10 & 20.68 & -10.78 \\ 0 & 0 & 6 & 0 & -9 & 1.38 & 46.83 & -58.14 & -4.53 \end{array} \right].$$

Set $k_0 = 0$, $i = 1$

$$L_0^{(0)} = \begin{bmatrix} 0 & 0 & 4 \\ 0 & 0 & 2 \\ 0 & 0 & 6 \end{bmatrix},$$

where $r_1 = \text{rank}(L_0^{(0)}) = 1 < (n = 3)$, $k_1 = n - r_1 = 2$.

The row block–matrix description of the c.s.p.m. of order $k_1 = 2$ is

$$U^{(1)} = \begin{bmatrix} 0 & 1 & 0 & | & 0 & 0 & 0 \\ 0 & 0 & 1 & | & 0 & 0 & 0 \\ 0 & 0 & 0 & | & 1 & 0 & 0 \end{bmatrix}.$$

Factorising $L_0^{(0)}$, using QR decomposition, we find $Q^{(1)} = I_n$. Now

$$L^{(1)} = L^{(0)}. < Q^{(1)}.U^{(1)} >_{\partial[L^{(0)}]}$$

$$= \begin{bmatrix} 4 & 1 & 2 & | & 11.96 & 71.03 & 57.38 & | & -24.65 & 0.00 & 0.00 \\ 2 & 1 & 1 & | & 4.62 & 32.10 & 20.68 & | & -10.78 & 0.00 & 0.00 \\ 6 & 0 & -9 & | & 1.38 & 46.83 & -58.14 & | & -4.53 & 0.00 & 0.00 \end{bmatrix}$$

Now, $i = 2$ and

$$L_0^{(1)} = \begin{bmatrix} 4 & 1 & 2 \\ 2 & 1 & 1 \\ 6 & 0 & -9 \end{bmatrix},$$

where $r_2 = \text{rank}(L_0^{(1)}) = 3 = n$. Hence, $L^{(1)}$ is now regular.

Set $\overline{L} = (L_0^{(1)})^{-1}.L^{(1)}$; i.e.,

$$\overline{L} = \begin{bmatrix} 1 & 0 & 0 & | & 2.81 & 16.55 & 11.34 & | & -5.39 & 0 & 0 \\ 0 & 1 & 0 & | & -2.72 & -6.83 & -16.02 & | & 3.09 & 0 & 0 \\ 0 & 0 & 1 & | & 1.72 & 5.83 & 14.02 & | & -3.09 & 0 & 0 \end{bmatrix}.$$

Thus, the companion matrix of $\overline{L}$ is

$$C_m(\overline{L}(q)) = \begin{bmatrix} -2.81 & -16.55 & -11.34 & 5.39 & 0 & 0 \\ 2.72 & 6.83 & 16.02 & -3.09 & 0 & 0 \\ -1.72 & -5.83 & -14.02 & 3.09 & 0 & 0 \\ 1 & 0 & 0 & 0 & 0 & 0 \\ 0 & 1 & 0 & 0 & 0 & 0 \\ 0 & 0 & 1 & 0 & 0 & 0 \end{bmatrix}.$$

Using the routine $POLY(\cdot)$, we find the characteristic polynomial of $C_m(\overline{L}(q))$; it is the coefficient vector

$$\overline{l} = POLY(C_m(\overline{L}(q))) = \begin{bmatrix} 1 & 10 & 37.97 & 67.01 & 79.96 & 0 & 0 \end{bmatrix}.$$

The relationship between the determinant polynomial of $L^{(1)}(q)$ and $L(q)$ is $\det(L^{(1)}(q)) = c.q^k.\det(L(q))$ or $l^{(1)}(q) = q^2.l(q)$, where $c = 1$. Therefore

$$l = DET(L_0^{(1)}).\begin{bmatrix} l_0^{(1)} & \dots & l_{\partial[l]-k}^{(1)} \end{bmatrix}$$

$$= \begin{bmatrix} -24 & -240 & -910.57 & -1600.71 & -1919.01 \end{bmatrix},$$

or, equivalently,

$$l(q) = -24q^4 - 240q^3 - 910.57q^2 - 1600.71q - 1919.01 \; .$$

2.4.8.1 Summary

At first sight, the algorithm and associated calculations may appear somewhat complicated and unattractive, but on closer examination we find that the operations involve only addition and multiplication of the coefficients of the polynomial matrix, and the factorisation of real matrices. The algorithm is well-conditioned due to the use of numerically robust routines for QR factorisation, $POLY(\cdot)$ and $DET(\cdot)$, which are provided by a software package.

2.4.9 Transforming a Full Row Rank Polynomial Matrix to its Row Proper Form

In the design process, described in Chapter 4, it is necessary to operate on polynomial matrices which are row–proper. However, in certain instances these polynomial matrices are not in row–proper form. In the following algorithm, which transforms a full rank polynomial matrix to its row–proper form, the calculations are simplified through the numerical representation of polynomial matrices.

Let us now consider the following definition and theorem found in Wolovich [118].

Definition 2.13 [118]: Two polynomial matrices $L_1(q) \in \Re[q]^{n_r \times n_c}$ and $L_2(q) \in \Re[q]^{n_r \times n_c}$ will be called *row equivalent* if and only if $L_1(q) = U(q).L_2(q)$, where $U(q)$ is a unimodular matrix.

Theorem 2.4 [118]: Any full rank polynomial matrix $L(q) \in \Re[q]^{n_r \times n_c}$ is row equivalent to a row–proper matrix; i.e., one can always find a $n_r \times n_r$ unimodular matrix $U(q)$ which reduces $L(q)$ to row–proper form.

Let us now consider the algorithm described by Wolovich [118].

Notation: The unimodular matrix $U(q)$ is composed of p factors:

$$U(q) = U^{(p)}(q) = \tilde{U}^{(p)}(q) \ldots \tilde{U}^{(2)}(q).\tilde{U}^{(1)}(q).\tilde{U}^{(0)}(q), \tag{2.15}$$

where $\tilde{U}^{(i)}(q) \in \Re[q]^{n_r \times n_r}$, $i \in 0, 1, \ldots, p$, is unimodular. Let $B(q) \in \Re[q]^{1 \times n_r}$ and $B(q) \not\equiv 0$. The kth polynomial element $B_k(q)$, which is to be determined, is a non–zero scalar.

Algorithm 2.4 Let us now consider the recursive algorithm which transforms $L(q)$ into its row–proper form.

Set $i = 0$, $L^{(0)}(q) = L(q)$ and $\tilde{U}^{(0)}(q) = I_{n_r}$ to start the algorithm.

Consider the ith iteration, $i \geq 1$, in the evaluation of $U(q)$ based on eqn. (2.15).

step 1. Construct the $n_r \times n_c$ polynomial matrix

$$P^{(i)}(q) = \text{diag}\left[\begin{array}{cccc} q^{\mu_1}, & q^{\mu_2}, & \ldots, & q^{\mu_{n_r}} \end{array}\right].\Gamma_r[L^{(i-1)}(q)]$$

of the polynomial matrix $L^{(i-1)}(q)$.

step 2. If $\text{rank}(\Gamma_r[L^{(i-1)}(q)]) = n_r$ the algorithm terminates and $U(q) = U^{(i-1)}(q)$, set $p = i - 1$.

step 3. Find a polynomial row vector $B(q)$ of minimal degree which satisfies

$$B(q).P^{(i)}(q) = 0 \tag{2.16}$$

with any polynomial element of $B(q)$, say the kth element $B_k(q)$, being a non–zero scalar.

step 4. Divide each element $B_j(q)$, $j \neq k$, of $B(q)$ by the non–zero scalar $B_k(q)$ such that $\tilde{B}(q) = B(q)/B_k(q)$.

step 5. The unimodular matrix $\tilde{U}^{(i)}(q)$ is the $n_r \times n_r$ identity matrix with the kth row of the identity matrix replaced by the polynomial row vector $\tilde{B}(q)$; i.e,

$$\tilde{U}^{(i)}(q) = \begin{bmatrix} 1 & 0 & \dots & 0 & 0 & \dots & 0 \\ 0 & 1 & & & & & \\ \vdots & \vdots & \ddots & & & & \\ \tilde{B}_1(q) & \tilde{B}_2(q) & & 1 & \tilde{B}_{k+1}(q) & \dots & \tilde{B}_{n_r}(q) \\ 0 & 0 & & & \ddots & & 0 \\ \vdots & \vdots & & & & \ddots & \\ 0 & 0 & & 0 & 0 & \dots & 1 \end{bmatrix} .$$

step 6. Set

$$U^{(i)}(q) = \tilde{U}^{(i)}(q).U^{(i-1)}(q) .$$

This ends the ith iteration.

Thus, the ith iteration of the algorithm results in

$$L^{(i)}(q) = \tilde{U}^{(i)}(q).L^{(i-1)}(q) .$$

$\nabla\nabla\nabla$

To illustrate the algorithm, let us consider the following example which makes use of the numerical techniques of polynomial matrices to realise a row–proper polynomial matrix.

Example 2.13 Let

$$L(q) = \begin{bmatrix} q^2-3 & 4q+2 & -q^2 \\ 1 & 2 & q+3 \\ 2q & 0 & -3q+2 \end{bmatrix},$$

the row block–matrix of which is

$$L = \left[\begin{array}{ccc|ccc|ccc} 1 & 0 & -1 & 0 & 4 & 0 & -3 & 2 & 0 \\ 0 & 0 & 0 & 0 & 0 & 1 & 1 & 2 & 3 \\ 0 & 0 & 0 & 2 & 0 & -3 & 0 & 0 & 2 \end{array}\right] .$$

Then the row block–matrix representation of the polynomial matrix $P^{(1)}(q)$ of $L(q)$ is

$$P^{(1)} = \left[\begin{array}{ccc|ccc|ccc} 1 & 0 & -1 & 0 & 0 & 0 & 0 & 0 & 0 \\ 0 & 0 & 0 & 0 & 0 & 1 & 0 & 0 & 0 \\ 0 & 0 & 0 & 2 & 0 & -3 & 0 & 0 & 0 \end{array}\right],$$

the highest row–degree coefficient matrix of the row block–matrix $L = L^{(0)}$ is

$$\Gamma_r[L^{(0)}] = \begin{bmatrix} 1 & 0 & -1 \\ 0 & 0 & 1 \\ 2 & 0 & -3 \end{bmatrix},$$

the row indices of which are $\mu_1 = 2$, $\mu_2 = 1$, $\mu_3 = 1$. Clearly, the row block–matrix $L^{(0)}$ is not row–proper since $\text{rank}(\Gamma_r[L^{(0)}]) \neq 3$.

We must now determine a polynomial row vector $B(q)$ of minimal degree which satisfies eqn. (2.16). This is achieved using the method of resultants to obtain a solution to the set of linear equations

$$B. < P^{(1)} >_{\partial[B]} = 0 \, .$$

Such a solution can be found by Gaussian elimination, say.

The row block–matrix

$$B = \begin{bmatrix} 0 & 1 & 1 & | & -2 & 0 & 0 \end{bmatrix}$$

or, equivalently,

$$B(q) = \begin{bmatrix} -2 & q & q \end{bmatrix}$$

represents an appropriate choice, where the non–zero scalar polynomial $B_1(q) = -2$. Dividing the row block–matrix B by the real scalar -2, we obtain

$$B = \begin{bmatrix} 0 & -0.5 & -0.5 & | & 1 & 0 & 0 \end{bmatrix}$$

or, equivalently,

$$B(q) = \begin{bmatrix} 1 & -0.5q & -0.5q \end{bmatrix}.$$

We can now proceed with the construction of the row block–matrix $\tilde{U}^{(1)}$; i.e,

$$\tilde{U}^{(1)} = \begin{bmatrix} 0 & -0.5 & -0.5 & | & 1 & 0 & 0 \\ 0 & 0 & 0 & | & 0 & 1 & 0 \\ 0 & 0 & 0 & | & 0 & 0 & 1 \end{bmatrix}$$

or, equivalently,

$$\tilde{U}^{(1)}(q) = \begin{bmatrix} 1 & -0.5q & -0.5q \\ 0 & 1 & 0 \\ 0 & 0 & 1 \end{bmatrix}.$$

Therefore the row block–matrix $L^{(1)}$ is

$$L^{(1)} = \tilde{U}^{(1)}. < L^{(0)} >_{\partial[\tilde{U}^{(1)}]} = \left[\begin{array}{ccc|ccc} -0.5 & 3 & -2.5 & -3 & 2 & 0 \\ 0 & 0 & 1 & 1 & 2 & 3 \\ 2 & 0 & -3 & 0 & 0 & 2 \end{array}\right]$$

or, equivalently,

$$L^{(1)}(q) = \tilde{U}^{(1)}(q).L^{(0)}(q) = \begin{bmatrix} -0.5q-3 & 3q+2 & -2.5q \\ 1 & 2 & q+3 \\ 2q & 0 & -3q+2 \end{bmatrix}.$$

The highest row–degree coefficient matrix of $L^{(1)}$ is

$$\Gamma_r[L^{(1)}] = \begin{bmatrix} -0.5 & 3 & -2.5 \\ 0 & 0 & 1 \\ 2 & 0 & -3 \end{bmatrix}$$

and is found to be of full–rank. Hence $L^{(1)}(q)$ is row–proper. Therefore $U^{(1)}(q) = \tilde{U}^{(1)}(q)$ is a unimodular matrix which reduces $L(q)$ to a row–proper form given by $L^{(1)}(q)$.

2.4.9.1 Summary

The only operations involved in the above methodology are addition and multiplication on the coefficients of a polynomial matrix and solving a set of linear equations by Gaussian elimination, say. The methodology is simple and numerically robust.

Chapter 3

Model Descriptions and Transformations Between Models

3.1 Introduction

In modelling systems or plants, a number of descriptions can be used to represent them. Two commonly–used model descriptions for linear, time–invariant, MIMO systems are the state–space description (SSD) and the transfer function matrix description (TFMD). However, for the design procedure proposed in Chapter 4, the models for both the plant and compensator will be represented by their matrix fraction description (MFD).

With reference to the diagram below, only three transformation algorithms are necessary to enable the transformation of any of the above–mentioned system models to any one of the remaining system models.

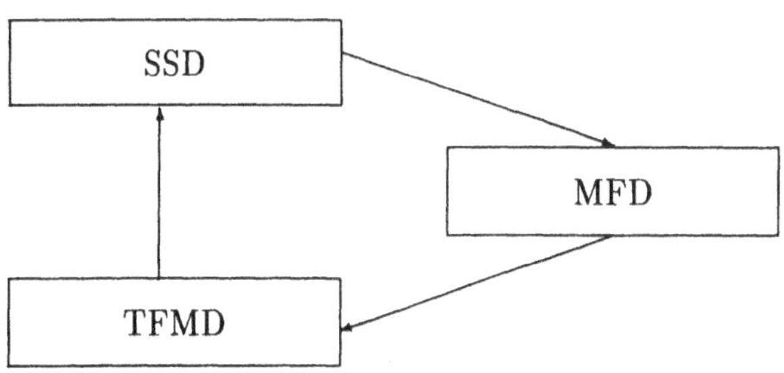

The following three transformations will be provided:

- SSD $\Rightarrow$ MFD;

- MFD $\Rightarrow$ TFMD; and
- TFMD $\Rightarrow$ SSD.

In section 3.2 of this chapter a summary of the properties of a MFD is presented. Then we derive in § 3.3 a computational algorithm which applies the techniques of § 2.4.4 and § 2.4.5 to realise the transformation SSD $\Rightarrow$ MFD. In order to realise a coprime MFD, the following are employed: (i) the concept of a minimal basis, (ii) the properties of the invariant indices of a SSD, (iii) the structure matrix and (iv) the skeleton block–matrix. In section 3.4, the methodologies of Appendix D and § 2.4.8 are combined to develop an algorithm which calculates the inverse of a square, non–singular polynomial matrix using its row block–matrix representation. By combining the methodology of § 2.4.5 and § 3.4 in section 3.5, an algorithm is realised which performs the transformation MFD $\Rightarrow$ TFMD. The transformation algorithm TFMD $\Rightarrow$ SSD is provided by a numerically robust function included in a software package [53]. Finally, in section 3.6 we consider the ARMA (AutoRegressive–Moving Average) model, which is used for the construction of the control algorithm for a digital compensator.

3.1.1 Mathematical Notation

The following notation will be used in the sections to follow. This chapter and the subsequent chapters consider both discrete– and continuous–time systems; to keep presentation compact and concise wherever possible no distinction is made between them. The time operator t denotes an instant in time. In the discrete–time case $t \in \mathcal{I}$ are integers, while for the continuous–time case $t \in \Re$ are the real numbers. Considerable use is made of the operator q. For example, the operations on a time varying signal $x(t)$ is described in the following way:

1. In the discrete–time domain the operator q represents the forward–difference operator, i.e.,

$$q.x(t) \triangleq x(t+1)\,.$$

The operator q is related to the Z–transformation as follows:

$$\mathcal{Z}\{q.x(t)\} = z.X(z) - z.x(0),$$

where $x(0)$ is a vector of initial conditions. The operator q can be directly replaced by the operator z if the initial conditions are assumed to be zero.

2. In the continuous–time domain the operator q represents the differential operator; i.e.,

$$q.x(t) \triangleq \dot{x}(t)\,.$$

The operator q is related to the Laplace transformation as follows:

$$\mathcal{L}\{q.x(t)\} = s.X(s) - x(0),$$

where $x(0)$ is a vector of initial conditions. The operator q can be directly replaced by the Laplace operator s if the initial conditions are assumed to be zero.

3.2 Properties of the MFD

3.2.1 Introduction

To describe the properties of the MFD, let us consider a quadruple of polynomial matrices, $(D(q),\ N_{\mathrm{L}}(q),\ N_{\mathrm{R}}(q),\ E(q))$, and represent the system by

$$D(q).\xi(t) = N_{\mathrm{L}}(q).u(t), \quad c(t) = N_{\mathrm{R}}(q).\xi(t) + E(q).u(t) \quad t \geq 0. \tag{3.1}$$

The system matrix is

$$S(q) = \begin{bmatrix} D(q) & N_{\mathrm{L}}(q) \\ -N_{\mathrm{R}}(q) & E(q) \end{bmatrix}, \tag{3.2}$$

such that

$$\begin{bmatrix} 0 \\ c(t) \end{bmatrix} = S(q).\begin{bmatrix} -\xi(t) \\ u(t) \end{bmatrix}. \tag{3.3}$$

This is called a polynomial matrix system description (PMSD) [102], in which:

- $D(q) \in \Re[q]^{n_\xi \times n_\xi}$, $N_{\mathrm{L}}(q) \in \Re[q]^{n_\xi \times n_i}$, $N_{\mathrm{R}}(q) \in \Re[q]^{n_o \times n_\xi}$, and $E(q) \in \Re[q]^{n_o \times n_i}$;

- $D(q)$ is non–singular;
- $\xi(t)$ is a n_ξ pseudo-state vector, $u(t)$ and $c(t)$ are n_i-input control and n_o-output response vectors of the PMSD;
- the characteristic polynomial, $d(q) \not\equiv 0$, of the PMSD is defined

$$d(q) \triangleq det(D(q)),$$

  the polynomial order of which is n_s;
- the order of the PMSD is the polynomial order, n_s, of the characteristic polynomial $d(q)$;
- the eigenvalues of the PMSD are any $\lambda \in \mathcal{C}$ such that $det(D(\lambda)) = 0$; and
- assuming zero initial conditions, the input–output behaviour of the PMSD is given by the transfer function

$$G(q) = N_R(q).D^{-1}(q).N_L(q) + E(q),$$

  where

$$c(t) = G(q).u(t)\,.$$

$\nabla\nabla\nabla$

The matrix fraction description (MFD) is a special case of a PMSD. The MFD can be represented in two forms. These being, (i) the right matrix fraction description (RMFD) and (ii) the left matrix fraction description (LMFD). Let us consider the properties of these two descriptions separately.

3.2.2 Properties of the RMFD

Definition 3.1 The general form of a RMFD is represented by the transfer function

$$G_R(q) = N_R(q).D_R^{-1}(q) + E(q) = (N_R(q) + E(q).D_R(q)).D_R^{-1}(q) = \overline{N}_R(q).D_R^{-1}(q) \quad (3.4)$$

the system matrix of which is

$$S(q) = \begin{bmatrix} D_{\mathrm{R}}(q) & I_{n_{\mathrm{i}}} \\ -N_{\mathrm{R}}(q) & E(q) \end{bmatrix}, \tag{3.5}$$

where $D_{\mathrm{R}}(q) \in \Re[q]^{n_{\mathrm{i}} \times n_{\mathrm{i}}}$ is the non-singular denominator, $\overline{N}_{\mathrm{R}}(q) \in \Re[q]^{n_{\mathrm{o}} \times n_{\mathrm{i}}}$ is the numerator, with $N_{\mathrm{R}}(q) \in \Re[q]^{n_{\mathrm{o}} \times n_{\mathrm{i}}}$ and $E(q) \in \Re[q]^{n_{\mathrm{o}} \times n_{\mathrm{i}}}$. The dimensions of the polynomial matrices are determined by the number of input control signals, n_{i}, and the number of output response signals, n_{o}, of the system or plant. The characteristic polynomial of the RMFD is $d(q) \triangleq det(D_{\mathrm{R}}(q))$, the polynomial order of which is n_{s}. The RMFD is equivalent to a completely controllable state–space system [45,102].

Definition 3.2 [16]: Let $A(q) = A_1(q).R(q) = L(q).A_2(q)$ and $B(q) = B_1(q).R(q) = L(q).B_2(q)$. The polynomial matrices $A(q)$ and $B(q)$ are said to be right–coprime (left–coprime) if and only if the greatest common right (left) divisor $R(q)$ ($L(q)$) of $A(q)$ and $B(q)$ is unimodular.

Definition 3.3 [16]: The zero $q_o \in C$ is called an output–decoupling zero of the RMFD if and only if, given any greatest common right divisor $R(q) \in \Re[q]^{n_{\mathrm{i}} \times n_{\mathrm{i}}}$ of $N_{\mathrm{R}}(q)$ and $D_{\mathrm{R}}(q)$, $det(R(q_o)) = 0$.

Comment: The output–decoupling zeros of a RMFD are equivalent to the unobservable modes of a state–space system.

Theorem 3.1 [107,16,102]: The RMFD of eqns. (3.4) and (3.5) is said to be right–coprime if and only if any one of the following equivalent conditions is satisfied.

1. The greatest common right divisor of $\overline{N}_{\mathrm{R}}(q)$ and $D_{\mathrm{R}}(q)$ is unimodular.

2. $\mathrm{rank}([D_{\mathrm{R}}^T(q)\ \overline{N}_{\mathrm{R}}^T(q)]) = n_{\mathrm{i}}$ for all $q \in C$.

3. There exist ([56] pp. 382) left–coprime polynomial matrices $R(q) \in \Re[q]^{n_{\mathrm{i}} \times n_{\mathrm{i}}}$ and $L(q) \in \Re[q]^{n_{\mathrm{i}} \times n_{\mathrm{o}}}$, such that

$$L(q).\overline{N}_{\mathrm{R}}(q) + R(q).D_{\mathrm{R}}(q) = I_{n_{\mathrm{i}}} .$$

4. The RMFD is equivalent to a completely observable state–space system.

5. The RMFD has no output–decoupling zeros.

6. The polynomial order of the characteristic polynomial, $d(q)$, of the RMFD is equal to the McMillan degree of $G_R(q)$.

Definition 3.4 [16,107,102]: The RMFD is said to be least order if and only if the RMFD is right–coprime.

Comment: With reference to Definition 3.4, a least order RMFD is equivalent to a completely controllable and observable SSD [102].

Theorem 3.2 [107,16]: The poles of the transfer function $G_R(q)$ are equal to the eigenvalues of its corresponding system matrix less all output–decoupling zeros; i.e.,

(eigenvalues of $D_R(q)$)=(output–decoupling zeros of RMFD)+ (poles of $G_R(q)$).

Because of system equivalence between the RMFD and SSD, the following definition is equivalent to Definition 3.4

Definition 3.5 The state–space system S(A,B,C,E) is completely observable if

$$\begin{bmatrix} qI_{n_s} - A \\ C \end{bmatrix}$$

are right–coprime; i.e., the observability matrix

$$\begin{bmatrix} C^T & A^T.C^T & \dots & (A^T)^{n_s-1}.C^T \end{bmatrix}$$

is of rank n_s.

Result 3.1 The dimension and degree of the denominator polynomial matrix and its characteristic polynomial are related through the inequality

$$n_i.\partial[D_R(q)] \geq \partial[d(q)]\,. \tag{3.6}$$

Result 3.2 For the system to satisfy causality, the plant or system is required to be proper (causal) or strictly proper (strictly causal); i.e.,

$$\partial[D_{\mathrm{R}}(q)] \geq \partial[\overline{N}_{\mathrm{R}}(q)]\,.$$

Furthermore, if $D_{\mathrm{R}}(q)$ is a column-proper polynomial matrix, then $\overline{N}_{\mathrm{R}}(q).D_{\mathrm{R}}^{-1}(q)$ is proper (causal) if and only if

$$\partial_{cj}[D_{\mathrm{R}}(q)] \geq \partial_{cj}[\overline{N}_{\mathrm{R}}(q)], j \in \underline{n_{\mathrm{i}}} \tag{3.7}$$

(see Wolovich [118]).

Definition 3.6 [56,15,102]: Let us now consider the invariants of a linear system (A, B), where $A \in \Re^{n_s \times n_s}$ and $B \in \Re^{n_s \times n_i}$ is a completely controllable pair. It has been reported by Brunovský [15], Wonham [125], Popov [98] and implicitly by Rosenbrock ([102], pp. 90) that there exists a list of integers $I_{\mathrm{r}}(A, B) \equiv \{\mu_1, \mu_2, \ldots, \mu_{n_{\mathrm{i}}}\}$ called the *controllability indices* of the couple (A, B) satisfying $0 \leq \mu_i \leq n_s$, $i \in \underline{n}_i$, and $\sum_{i=1}^{n_{\mathrm{i}}} \mu_i = n_s$. This set of indices is a complete set of invariants of a controllable couple (A, B). Moreover, the indices $I_{\mathrm{r}}(A, B)$ completely define the structure of the couple (A, B) [59,98] and is invariant under state coordinate transformation; i.e., $I_{\mathrm{r}}(A, B) = I_{\mathrm{r}}(P.A.P^{-1},\ P.B)$, for any non-singular matrix $P \in \Re^{n_s \times n_s}$. Appendix E contains a methodology, found in Kailath [56], which computes the controllability indices $I_{\mathrm{r}}(A, B)$ of a completely controllable state-space system $S(A, B, C, E)$. For further details on invariants of linear systems, see [59, 98,102].

Definition 3.7 [115]: Let $G(q)$ be a $n_{\mathrm{o}} \times n_{\mathrm{i}}$ proper (causal) rational matrix and let $S(A, B, C, E)$ be any controllable and observable realisation of $G(q)$, (i.e., $G(q) = C.(qI - A)^{-1}.B + E = N_{\mathrm{R}}(q).D_{\mathrm{R}}^{-1}(q)$ where A is of least possible size). A set of controllability indices of $G(q)$ is defined to be equal to $I_{\mathrm{r}}(A, B)$ – the set of controllability indices of the couple (A, B).

The controllability indices of Definitions 3.6 and 3.7 are used in § 3.3 to realise a right-coprime RMFD.

3.2.3 Properties of the LMFD

The left matrix fraction description has similar mathematical properties to that of the right matrix fraction description.

Definition 3.8 The general form of a LMFD is represented by the transfer function

$$G_L(q) = D_L^{-1}(q).N_L(q) + E(q) = D_L^{-1}(q).(N_L(q) + D_L(q).E(q)) = D_L^{-1}(q).\overline{N}_L(q), \quad (3.8)$$

the system matrix of which is

$$S(q) = \begin{bmatrix} D_L(q) & N_L(q) \\ -I_{n_o} & E(q) \end{bmatrix}, \quad (3.9)$$

where $D_L(q) \in \Re[q]^{n_o \times n_o}$ is the non-singular denominator, $\overline{N}_L(q) \in \Re[q]^{n_o \times n_i}$ is the numerator, with $N_L(q) \in \Re[q]^{n_o \times n_i}$ and $E(q) \in \Re[q]^{n_o \times n_i}$. The dimensions of the polynomial matrices are determined by the number of input control signals, n_i, and the number of output response signals, n_o, of the system or plant. The characteristic polynomial of the LMFD is $d(q) \triangleq det(D_L(q))$, the polynomial order of which is n_s. The LMFD is equivalent to a completely observable state-space system [45,102].

Definition 3.9 [16]: The zero $q_i \in C$ is called an input-decoupling zero of the LMFD if and only if, given any greatest common left divisor $L(q) \in \Re[q]^{n_o \times n_o}$ of $N_L(q)$ and $D_L(q)$, $det(L(q_i)) = 0$.

Comment: The input-decoupling zeros of a LMFD are equivalent to uncontrollable modes of a state-space system.

Theorem 3.3 [107,16,102]: The LMFD of eqns. (3.8) and (3.9) is said to be left coprime if and only if any one of the following equivalent conditions is satisfied.

1. The greatest common left divisor of $\overline{N}_L(q)$ and $D_L(q)$ is unimodular.

2. $\text{rank}([D_L(q) \ \overline{N}_L(q)]) = n_o$ for all $q \in C$.

3. There exist right-coprime polynomial matrices $R(q) \in \Re[q]^{n_o \times n_o}$ and $L(q) \in \Re[q]^{n_i \times n_o}$, such that

$$\overline{N}_L(q).L(q) + D_L(q).R(q) = I_{n_o}.$$

4. The LMFD is equivalent to a completely controllable state–space system.

5. The LMFD has no input–decoupling zeros.

6. The polynomial order of the characteristic polynomial, $d(q)$, of the LMFD is equal to the McMillan degree of $G_{\mathrm{L}}(q)$.

Definition 3.10 [16,107,102]: The LMFD is said to be least order if and only if the LMFD is left–coprime.

Comment: With reference to Definition 3.10, a least order LMFD is equivalent to a completely controllable and observable SSD [102].

Theorem 3.4 [107,16]: The poles of the transfer function $G_{\mathrm{L}}(q)$ are equal to the eigenvalues of its corresponding system matrix less all input–decoupling zeros; i.e.,

(eigenvalues of $D_{\mathrm{L}}(q)$)=(input–decoupling zeros of LMFD)+(poles of $G_{\mathrm{L}}(q)$).

Because of system equivalence between the LMFD and SSD, the following definition is equivalent to Definition 3.10.

Definition 3.11 The state–space system S(A,B,C,E) is completely controllable if

$$\begin{bmatrix} qI_{n_s} - A & -B \end{bmatrix}$$

are left–coprime; i.e., the controllability matrix

$$\begin{bmatrix} B & A.B & \ldots & A^{n_s-1}.B \end{bmatrix}$$

is of rank n_s.

Result 3.3 The degree and dimension of the denominator polynomial matrix and its characteristic polynomial are related through the inequality

$$n_o.\partial[D_{\mathrm{L}}(q)] \geq \partial[d(q)]\,. \tag{3.10}$$

Result 3.4 For the system to satisfy causality, the plant or system is required to be proper (causal) or strictly proper (strictly causal); i.e.,

$$\partial[D_{\mathrm{L}}(q)] \geq \partial[\overline{N}_{\mathrm{L}}(q)]\,.$$

Furthermore, if $D_{\mathrm{L}}(q)$ is a row-proper polynomial matrix, then $D_{\mathrm{L}}^{-1}(q).N_{\mathrm{L}}(q)$ is proper (causal) if and only if

$$\partial_{\mathrm{r}i}[D_{\mathrm{L}}(q)] \geq \partial_{\mathrm{r}i}[\overline{N}_{\mathrm{L}}(q)], i \in \underline{n_{\mathrm{o}}} \tag{3.11}$$

(see Wolovich [118]).

Definition 3.12 [115,15,102]: Let $C \in \Re^{n_o \times n_s}$ and $A \in \Re^{n_s \times n_s}$ be a completely observable couple. There exists a list of integers $I_l(C, A) \equiv \{\nu_1,\ \nu_2,\ \ldots,\ \nu_{n_o}\}$ called the *observability indices* of the couple (C, A) satisfying $0 \leq \nu_j \leq n_s$, $j \in \underline{n}_o$, and $\sum_{j=1}^{n_o} \nu_j = n_s$. This set of indices is a complete set of invariants of an observable couple (C, A). Moreover, the invariants $I_l(C, A)$ completely defines the structure of the couple (C, A) and is invariant under state coordinate transformation; i.e., $I_l(C, A) = I_l(C.P, P.A.P^{-1})$, for any non-singular matrix $P \in \Re^{n_s \times n_s}$. A methodology similar to that found in Appendix E can be used to realise the observability indices $I_l(C, A)$ of a completely observable state-space system $S(A, B, C, E)$, where the controllability matrix is replaced by the observability matrix.

Definition 3.13 Let $G(q)$ be a $n_{\mathrm{o}} \times n_{\mathrm{i}}$ proper (causal) rational matrix and let $S(A, B, C, E)$ be any controllable and observable realisation of $G(q)$, (i.e., $G(q) = C.(qI - A)^{-1}.B + E = D_{\mathrm{L}}^{-1}(q).N_{\mathrm{L}}(q)$ where A is of least possible size). A set of observability indices of $G(q)$ is defined to be equal to $I_l(C, A)$ – the set of observability indices of the couple (C, A).

The right–coprime RMFD or left–coprime LMFD described above can be regarded as extensions to the multivariable case of the classical single–input, single–output transfer functions with coprime numerator and denominator polynomials.

3.3 Numerical Algorithms for the Transformation from SSD to MFD

3.3.1 Introduction

To perform the system transformations requires the development of computational algorithms which transform a SSD or TFMD to its MFD. The algorithms which are discussed in this section involve only the transformation SSD $\Rightarrow$ MFD. The transformation TFMD $\Rightarrow$ MFD is performed by combining the transformations TFMD $\Rightarrow$ SSD (provided in commercial software packages) and SSD $\Rightarrow$ MFD.

It is well known [102,118] that a rational function $G(q)$ can be represented by the following MFD:

$$G(q) = N_{\mathrm{R}}(q).D_{\mathrm{R}}^{-1}(q) = D_{\mathrm{L}}^{-1}(q).N_{\mathrm{L}}(q) \tag{3.12}$$

where $N_{\mathrm{R}}(q)$, $D_{\mathrm{R}}(q)$, are right–coprime and $N_{\mathrm{L}}(q)$, $D_{\mathrm{L}}(q)$ are left–coprime polynomial matrices with $\partial[N_{\mathrm{R}}(q)] \leq \partial[D_{\mathrm{R}}(q)]$ and $\partial[N_{\mathrm{L}}(q)] \leq \partial[D_{\mathrm{L}}(q)]$. In recent years, MFDs have been used widely in linear control theory, e.g., to define poles and zeros of transfer function matrices [25], to obtain controllable and observable state–space realisations of transfer functions [119], and to find proper (causal) rational function matrices $G(q)$ by solving equations of the form $G(q).D_{\mathrm{R}}(q) = N_{\mathrm{R}}(q)$, where $N_{\mathrm{R}}(q)$ and $D_{\mathrm{R}}(q)$ are matrices of polynomials [64,115].

It is therefore not surprising that several methods [40,42,45,64,105,115,120] are available for obtaining MFDs. In general, it is not difficult to construct a rational function matrix as a product of a polynomial matrix and the inverse of another polynomial matrix. The difficulty arises when one is interested in a coprime (least order) MFD. In engineering applications, the underlying system can be assumed finite and it is often important to obtain a least–order model, particularly if the model is to be used to design a compensator. An approach used by Rosenbrock [102] requires elementary row operations on the matrix $[G(q) \;\; - I]$, which involves multiplication and division by polynomials. The method is numerically unstable and is difficult to implement on a computer. In [40,42,64,105,115] the method of determining a proper (causal) rational function matrix searches for a minimal polynomial basis [40] for the right or left null space of a polynomial matrix. The minimal polynomial basis is particularly relevant to our methodology in the determination of a coprime MFD. The methods in [40,42,64,105,115] can be used to determine such a basis. However, in [40,105,115] the emphasis is on developing the theory, and the numerical properties of the proposed methods are not considered with the result that the methods are computationally involved and prone to numerical difficulties. In contrast, in [42,45,64] consideration has been given to developing numerically robust computational methods which are more attractive for computer software development. The main disadvantage of [42,64], however, is that they involve operations on matrices, the orders of which may be much higher than the dimension of the state–space system.

In contrast, it is proposed to combine (i) the methodology presented by Gibbard and Papliński [45], (ii) the ideas of a minimal basis presented by Forney [40], (iii) the invariant controllability/observability indices of a SSD [56,98], (iv) the structure matrix and (v) the skeleton block–matrix to realise an algorithm which is easily implemented by computer software. Such an algorithm directly performs the transformation SSD $\Rightarrow$ MFD to obtain a co–prime MFD.

The methodology proposed in [45] is formulated in terms of exact model matching; i.e.,

$$G(q) = C.D^{-1}(q).B + E = N_{\rm R}(q).D_{\rm R}^{-1}(q) = D_{\rm L}^{-1}(q).N_{\rm L}(q),$$

and makes use of the method of resultants to transform the problem into a set of linear equations. The resulting solution describes the denominator and numerator polynomial matrices of the RMFD (LMFD) by their column (row) block–matrix descriptions.

Due to the similarity of the methodologies used in realising the transformations from SSD to either RMFD or LMFD, a complete analysis will only be given for the transformation SSD $\Rightarrow$ RMFD.

3.3.2 An Algorithm to Transform a SSD to RMFD

Let us consider, in detail, the methodology developed in [45].

Proposition 3.1 Let the linear time–invariant system be described by the SSD

$$\dot{x}(t) = A.x(t) + B.u(t)$$

$$c(t) = C.x(t) + E.u(t),$$

where $x(t) \in \Re^{n_s}$, $u(t) \in \Re^{n_i}$ and $c(t) \in \Re^{n_o}$, and

$A \in \Re^{n_s \times n_s}$ is the plant matrix,

$B \in \Re^{n_s \times n_i}$ is the input matrix,

$C \in \Re^{n_o \times n_s}$ is the output matrix and

$E \in \Re^{n_o \times n_i}$ is the transmission matrix.

Assumption 3.1 The state–space system S(A,B,C,E) is completely controllable and observable.

$\nabla\nabla\nabla$

Let the system transfer function of the state–space model be described by

$$G_{SS}(q) = G_S(q) + E \tag{3.13}$$

$$= C.D^{-1}(q).B + E,$$

where

$$D(q) = qI_{n_s} - A\,.$$

Let the RMFD be described by

$$G_{RR}(q) = G_R(q) + E$$

$$= N_R(q).D_R^{-1}(q) + E$$

$$= [N_R(q) + E.D_R(q)].D_R^{-1}(q)\,.$$

Equating the system transfer functions $G_{SS}(q)$ and $G_{RR}(q)$, we obtain

$$G_S(q) + E = G_R(q) + E$$

or, equivalently,

$$G_S(q) = G_R(q)\,.$$

Thus

$$C.D^{-1}(q).B = N_R(q).D_R^{-1}(q)\,.$$

Let $N_B(q)$ denote a polynomial matrix which satisfies the following expression

$$D^{-1}(q).B = N_B(q).D_R^{-1}(q), \tag{3.14}$$

where

$$N_R(q) = C.N_B(q)\,. \tag{3.15}$$

Rearranging eqn. (3.14) by pre-multiplying by $D(q)$ and post-multiplying by $D_R(q)$, we obtain

$$B.D_R(q) = D(q).N_B(q) \tag{3.16}$$

or, equivalently,

$$\begin{bmatrix} B & -D(q) \end{bmatrix} \cdot \begin{bmatrix} D_R(q) \\ N_B(q) \end{bmatrix} = 0\,. \tag{3.17}$$

Equation (3.17) can now be converted to its numerical equivalent by using the appropriate polynomial block-matrices and resultants (see § 2.4.5). Thus, eqn. (3.17) can be rewritten as a set of linear equations, for which the column block-matrix descriptions of $D_R(q)$ and $N_B(q)$ are to be determined; i.e.,

$$\begin{bmatrix} <B>^b_{\partial[D^b_R]} & <D>^b_{\partial[N^b_B]} \end{bmatrix} \cdot \begin{bmatrix} -D^b_R \\ N^b_B \end{bmatrix} = 0 \tag{3.18}$$

or, equivalently,

$$F \qquad\qquad Y \;= 0\,.$$

The transfer function $G_S(q)$ of eqn. (3.13) is strictly proper (causal). Moreover, $G_R(q)$, which represents the RMFD of $G_S(q)$, is also strictly proper (causal); i.e., $\partial[N^b_R] \leq \partial[N^b_B] < \partial[D^b_R]$. Therefore, the greatest degree which $N_B(q)$ can have is $\partial[N^b_B] = \partial[D^b_R] - 1$. The exact degree of $N_B(q)$ is not known, so it is assumed to be of degree $\partial[N_B^b] = \partial[D_R^b] - 1$. Using the maximum possible degree for N_B^b some of the most significant coefficient block-matrices of the solution to the column block-matrix N_B^b may be zero matrices.

With reference to eqn. (3.18):

D_R^b is the unknown denominator column block-matrix

$$D_R^b = \begin{bmatrix} D_{R_0} & \dots & D_{R_{\partial[D_R^b]}} \end{bmatrix}^b$$

of the RMFD, the dimension of which is $n_i.(\partial[D_R^b] + 1) \times n_i$.

N_B^b is the unknown column block-matrix

$$N_B^b = \begin{bmatrix} N_{B0} & \dots & N_{B\partial[D_R^b]-1} \end{bmatrix}^b,$$

the dimension of which is $n_s.(\partial[N_B^b]+1) \times n_i$, and is used to determine the numerator column block-matrix N_R^b of the RMFD.

$< B >^b_{\partial[D_R^b]}$ is the block-transpose of the resultant of order $\partial[D_R^b]$ of the zero-degree polynomial matrix B. The dimension of $< B >^b_{\partial[D_R^b]}$ is $n_s.(\partial[D_R^b]+1) \times n_i.(\partial[D_R^b]+1)$.

$< D >^b_{\partial[D_R^b]-1}$ is the block-transpose of the resultant of order $\partial[D_R^b] - 1$ of the first degree monic polynomial matrix $D(q) = q.I_{n_s} - A$ or $D = [I_{n_s} \mid -A]$. The dimension of $< D >^b_{\partial[D_R^b]-1}$ is $n_s.(\partial[D_R^b] + 1) \times n_s.\partial[D_R^b]$.

F is the matrix of resultants $[< B >^b_{\partial[D_R^b]} < D >^b_{\partial[D_R^b]-1}]$.

Y is the matrix of unknowns

$$\begin{bmatrix} -D_R^b \\ N_B^b \end{bmatrix}.$$

The relationship described by eqn. (3.18) now consists of a set of $n_s.(\partial[D_R^b] + 1)$ linear equations, the unknowns of which are the block-matrices D_R^b and N_B^b. There are two possible outcomes to the solution set of the linear equations (3.18). These being, (i) a non-zero parametric solution and (ii) a homogeneous solution. The only desirable outcome is the non-zero parametric solution which requires the number of unknowns to be greater than the number of linearly independent linear equations of eqn. (3.18). In equivalent terms, the dimension of the matrix F must be such that the number of columns is greater than the rank of the matrix F; i.e., $n_c(F) > \text{rank}(F)$. The row and column dimensions of the matrix F are, respectively,

$$n_r(F) = (\partial[D_R^b] + 1).n_s \text{ and } n_c(F) = (\partial[D_R^b] + 1).n_i + \partial[D_R^b].n_s\,.$$

To illustrate the structure of the matrix F let us assume that the degree $\partial[D_R^b] = 2$, hence

$$\begin{bmatrix} B & 0 & 0 & | & I_{n_s} & 0 \\ 0 & B & 0 & | & -A & I_{n_s} \\ 0 & 0 & B & | & 0 & -A \end{bmatrix} \cdot \begin{bmatrix} -D_{R_0} \\ -D_{R_1} \\ -D_{R_2} \\ N_{B_0} \\ N_{B_1} \end{bmatrix} = 0\,. \tag{3.19}$$

Let us premultiply eqn. (3.18) by the inverse of the lower-triangular Toeplitz block-matrix of D^b of order $\partial[D_R^b]$ (see Appendix B). It can be shown that eqn. (3.18) becomes

$$\begin{bmatrix} TL^{-1}_{\partial[D_R^b]}(D^b).<B>^b_{\partial[D_R^b]} & \begin{array}{c|c} & I_{n_s.\partial[D_R^b]} \\ | - & - \\ & 0_{n_s \times n_s.\partial[D_R^b]} \end{array} \end{bmatrix} \cdot \begin{bmatrix} -D_R^b \\ N_B^b \end{bmatrix} = 0, \tag{3.20}$$

using the relationship

$$TL^{-1}_{\partial[D_R^b]}(D^b) = TL^{-1}_{\partial[D_R^b]}\left(\begin{bmatrix} I_{n_s} \\ -A \end{bmatrix}\right) = TL_{\partial[D_R^b]}\left(\begin{bmatrix} I_{n_s} \\ A \\ \vdots \\ A^{\partial[D_R^b]} \end{bmatrix}\right),$$

where $I_{n_i.\partial[D_R^b]}$ denotes a identity matrix of dimension $n_i.\partial[D_R^b]$ and $0_{n_s \times n_s.\partial[D_R^b]}$ denotes a $n_s \times n_s.\partial[D_R^b]$ zero matrix.

Let us replace

$$TL^{-1}_{\partial[D_R^b]}(D^b).<B>^b_{\partial[D_R^b]}$$

by the lower-triangular Toeplitz matrix

$$TL_{\partial[D_R^b]}(CT^b_{\partial[D_R^b]}),$$

where

$$CT_j \triangleq [B\ A.B\ A^2.B\ \ldots\ A^j.B]$$

is defined as the controllability matrix of order j. Thus eqn. (3.20) can now be written as

$$\left[\begin{array}{c|c} TL_{\partial[D_R^b]}(\mathcal{CT}^b_{\partial[D_R^b]}) & \begin{array}{c} I_{n_s.\partial[D_R^b]} \\ - \\ 0_{n_s \times n_s.\partial[D_R^b]} \end{array} \end{array} \right] . \left[\begin{array}{c} -D_R^b \\ N_B^b \end{array} \right] = 0 . \tag{3.21}$$

Again, to illustrate the structure of eqn. (3.21) let the degree $\partial[D_R^b] = 2$, hence

$$\left[\begin{array}{ccc|cc} B & 0 & 0 & I_{n_s} & 0 \\ A.B & B & 0 & 0 & I_{n_s} \\ & & & - & - \\ A^2.B & A.B & B & 0 & 0 \end{array} \right] . \left[\begin{array}{c} -D_{R_0} \\ -D_{R_1} \\ -D_{R_2} \\ N_{B_0} \\ N_{B_1} \end{array} \right] = 0 \tag{3.22}$$

or

$$F \qquad\qquad Y \quad = 0.$$

The unknowns D_R^b and N_B^b can be partially separated into two equations. The two equations can be written as follows.

Premultiplying $[-D_R \; N_B]^b$,

(i) by block row $\partial[D_R^b] + 1$ of F, we have

$$\mathcal{AB}_{\partial[D_R^b]}.D_R^b = 0, \tag{3.23}$$

where

$$\mathcal{AB}_{\partial[D_R^b]} = \left[\begin{array}{cccc} A^{\partial[D_R]}.B & \dots & A.B & B \end{array} \right] \tag{3.24}$$

is the controllability matrix of order $\partial[D_R^b]$ with the sequence of the block elements reversed;

(ii) by block rows 1 to $\partial[D_R^b]$ of F, we have

$$N_B^b = TL_{\partial[D_R^b]-1}(\mathcal{CT}^b_{\partial[D_R^b]-1}).\tilde{D}_R^b, \tag{3.25}$$

where

$$\mathcal{CT}_{\partial[D_R^b]-1} = \left[\begin{array}{cccc} B & A.B & \dots & A^{\partial[D_R^b]-1}.B \end{array} \right]$$

is the controllability matrix of order $\partial[D_R^b] - 1$, and

$$\tilde{D}_R^b = \left[\begin{array}{c} D_{R_0} \\ \vdots \\ D_{R_{\partial[D_R^b]-1}} \end{array} \right] .$$

Assuming a solution for D_{R}^{b} from eqn. (3.23) and following the calculation of N_{B}, we can determine the numerator block–matrix N_{R} using eqn. (3.15); i.e.,

$$N_{\mathrm{R}} = C.N_{\mathrm{B}}\,.$$

Substitution of eqn. (3.25) into the above equation yields the the following expression for the coefficients of the numerator polynomial matrix $N_{\mathrm{R}}(q)$.

$$N_{\mathrm{R}}^{b} = TL_{\partial[D_{\mathrm{R}}^{b}]-1}(M_{\partial[D_{\mathrm{R}}^{b}]-1}^{b}).\tilde{D}_{\mathrm{R}}^{b}, \tag{3.26}$$

where

$$M_j \triangleq \left[\begin{array}{ccccc} C.B & C.A.B & C.A^2.B & \ldots & C.A^j.B \end{array}\right]$$

is known as the matrix of Markov parameters.

If the transmission matrix E is a non–zero matrix then the numerator is modified to

$$\left[\begin{array}{c} 0_{n_o \times n_i} \\ N_{\mathrm{R}}^{b} \end{array}\right] + < E >_{\partial[D_{\mathrm{R}}^{b}]}^{b} .D_{\mathrm{R}}^{b}, \tag{3.27}$$

where $0_{n_o \times n_i}$ is the $n_o \times n_i$ matrix of zeros.

Note that a similar methodology can be used to realise the transformation SSD⇒ LMFD [45]. To calculate the denominator row block–matrix, D_{L}, and numerator row block–matrix, N_{L}, of the LMFD, the following equations are employed:

$$D_{\mathrm{L}}.\mathcal{CA}_{\partial[D_{\mathrm{L}}]} = 0,$$

where

$$\mathcal{CA}_j = \left[\begin{array}{c} CA^j \\ \vdots \\ CA \\ C \end{array}\right]$$

is the observability matrix of order j with the sequence of the block elements reversed, and

$$N_{\mathrm{L}} = \tilde{D}_{\mathrm{L}}.TU_{\partial[D_{\mathrm{L}}]-1}(M_{\partial[D_{\mathrm{L}}]-1})$$

where

$$\tilde{D}_{\mathrm{L}} = \left[\begin{array}{ccc} D_{\mathrm{L}_0} & \ldots & D_{\mathrm{L}_{\partial[D_{\mathrm{L}}]-1}} \end{array}\right]$$

and $TU_j(\cdot)$ represents the upper–triangular Toeplitz block–matrix of order j.

In order to perform the transformation SSD⇒ RMFD, two equations, (3.23) and (3.26), are required, i.e.

$$\mathcal{AB}_{\partial[D_R^b]}.D_R^b = 0 \tag{3.28}$$

and

$$N_R^b = TL_{\partial[D_R^b]-1}(M^b_{\partial[D_R^b]-1}).\tilde{D}_R^b\,. \tag{3.29}$$

In order to obtain a solution to the denominator and numerator column block–matrices D_R^b and N_R^b, we must first solve eqn. (3.28) for the denominator column block–matrix, D_R^b; the numerator column block–matrix, N_R^b, can then be calculated from eqn. (3.29).

Let us, therefore, consider eqn. (3.28). To derive a solution for D_R^b we must search for a basis, D_R^b, of column dimension n_i which spans a subspace of the null space of $\mathcal{AB}_{\partial[D_R^b]}$, such that the resulting polynomial matrix description, $D_R(q)$, of D_R^b is of least order. Moreover, the denominator $D_R(q)$ is a minimal basis of order n_s such that $N_R(q).D_R^{-1}(q)$ is right–coprime, and hence of least order. We now need the following lemmas.

Lemma 3.1 [57]: The McMillan degree of a $n_o \times n_i$, coprime, proper(causal) rational matrix $G(q)$ is the sum of the controllability (observability) indices of $G(q)$; i.e.,

$$n_s = \sum_{i=1}^{n_i} \mu_i = (\sum_{j=1}^{n_o} \nu_j)$$

Lemma 3.2 Let $D_R(q)$ be a row–proper matrix, the row structure matrix of which is

$$P_r(q) = diag\left[\; q^{\mu_1}, \quad q^{\mu_2}, \quad \ldots, \quad q^{\mu_{n_i}} \;\right] \cdot \Gamma_r[D_R(q)]\,. \tag{3.30}$$

The RMFD, $N_R(q).D_R^{-1}(q)$, is right–coprime (least order) if and only if the indices μ_j, $j \in \underline{n}_i$, are the controllability indices of $G_R(q)$ (see Definition 3.7).

Proof 3.1 The result follows immediately from a definition equivalent to that of Definition 2.11, item 6, for a row structure matrix, Definitions 3.6 and 3.7 of the controllability indices, item 6 of Theorem 3.1 and Lemma 3.1.

∇∇∇

The following proposition states that a consistent solution to the set of linear equations (3.28) can always be determined by solving for the n_s unknown residue rows of the skeleton column block–matrix, $D^b_{\mathrm{R}}s$, of the denominator polynomial matrix, $D_{\mathrm{R}}(q)$, for which the row structure matrix $P_{\mathrm{r}}(q)$ is given in eqn. (3.30).

Proposition 3.2 Let $D^b_{\mathrm{R}}s$ denote the skeleton column block–matrix of the row structure matrix, $P_{\mathrm{r}}(q)$, as is described by eqn. (3.30). Let

$$\overline{D}^b_{\mathrm{R}}s = \begin{bmatrix} Ps \\ Es \end{bmatrix}$$

and

$$\overline{\mathcal{AB}}_{\partial[D^b_{\mathrm{R}}]} = \begin{bmatrix} \mathcal{AB}_{\mathrm{Ps}} & \mathcal{AB}_{\mathrm{Es}} \end{bmatrix}.$$

If necessary, both row reordering of $D^b_{\mathrm{R}}s$ and column reordering of $\mathcal{AB}_{\partial[D^b_{\mathrm{R}}]}$ are carried out so that Ps contains both the n_{i} active structure rows and $(n_{\mathrm{i}}.\partial[D^b_{\mathrm{R}}]-n_s)$ inactive structure rows, and Es contains the n_s residue rows of the skeleton column block–matrix. Note that $\mathcal{AB}_{\mathrm{Ps}}$ and $\mathcal{AB}_{\mathrm{Es}}$ are the matrices, the columns of which correspond to the rows of the matrices Ps and Es, respectively, such that $\overline{\mathcal{AB}}_{\partial[D^b_{\mathrm{R}}]}.\overline{D}^b_{\mathrm{R}}s = 0$. Then the consistency rank condition

$$\mathrm{rank}\begin{bmatrix} \mathcal{AB}_{\mathrm{Ps}} & | & \mathcal{AB}_{\mathrm{Es}} \end{bmatrix} = \mathrm{rank}\,[\mathcal{AB}_{\mathrm{Es}}] = n_s \tag{3.31}$$

is satisfied.

Proof 3.2 With reference to the methodology described in Appendix E, which computes the controllability indices of a completely controllable SSD, the n_s columns of the matrix $\mathcal{AB}_{\mathrm{Es}}$, i.e. the columns

$$b_1,\quad A.b_1,\quad \ldots,\quad A^{\mu_1-1}.b_1,\quad b_2,\quad A.b_2,\quad \ldots,\quad A^{\mu_{n_{\mathrm{i}}}-1}.b_{n_{\mathrm{i}}},$$

which correspond to the n_s residue rows of the column block–matrix $D^b_{\mathrm{R}}s$ (or, equivalently, the rows of the matrix Es), are linearly independent. Hence, the consistency rank condition given by eqn. (3.31) is satisfied.

$\nabla\nabla\nabla$

Algorithm 3.1 The following steps summarise the algorithm which transforms a state-space description to its right matrix fraction description.

step 1. Compute the controllability indices, $I_r(A,B)$, of the state-space system $S(A,B,C,E)$ using the methodology described in Appendix E.

step 2. Construct the row structure matrix

$$P_r(q) = diag\left[\begin{array}{cccc} q^{\mu_1}, & q^{\mu_2}, & \ldots, & q^{\mu_{n_i}} \end{array}\right] \cdot (\Gamma_r[D_R(q)] = I_{n_i})$$

where the highest row-degree coefficient matrix $\Gamma_r[D_R(q)]$ of $D_R(q)$ is set to the $n_i \times n_i$ identity matrix, without loss of generality.

step 3. Construct the skeleton column block-matrix, $D_R^b s$, of $D_R(q)$ using the row structure matrix, $P_r(q)$, of step 2.

step 4. Solve for the residue rows of the skeleton column block-matrix, $D_R s$, of $D_R(q)$ using the equation

$$\mathcal{AB}_{\partial[D_R]} . D_R^b s = 0 \,.$$

step 5. Calculate the numerator

$$N_R^b = TL_{\partial[D_R]-1}(M^b_{[D_R]-1}).\tilde{D}_R^b \,.$$

step 6. If the transmission matrix, E, of the SSD is not a zero matrix then the numerator is modified to

$$\left[\begin{array}{c} 0_{n_o \times n_i} \\ N_R^b \end{array}\right] + < E >^b_{\partial[D_R^b]} .D_R^b \,.$$

▽▽▽

To illustrate the algorithm, let us consider the following two examples. The computations are performed with the aid of a software package using double precision arithmetic.

Example 3.1 Consider a controllable and observable continuous state-space system with $n_s = 4$ states, $n_i = 2$ inputs and $n_o = 3$ outputs.

$$A = \left[\begin{array}{cccc} -1 & 0 & 3 & -1 \\ 4 & -2 & 1 & 2 \\ 3 & 0 & -3 & 4 \\ -4 & -2 & -3 & -4 \end{array}\right], B = \left[\begin{array}{cc} 2 & 1 \\ 0 & 2 \\ 1 & -1 \\ -1 & 0 \end{array}\right],$$

$$C = \begin{bmatrix} -1 & 2 & 1 & -2 \\ 0 & -1 & -2 & 0 \\ 1 & 1 & 0 & -2 \end{bmatrix} \text{ and } E = \begin{bmatrix} 0 & 0 \\ 0 & 0 \\ 0 & 0 \end{bmatrix}.$$

step 1. Using the methodology described in Appendix E the set of controllability indices of the state-space system is $I_r(A, B) = \{\mu_1 = 2,\ \mu_2 = 2\}$.

step 2. The row structure matrix of $D_R(s)$ is:

$$P_r(s) = diag\left[\ s^{\mu_1=2},\quad s^{\mu_2=2}\ \right].I_2$$

or, equivalently,

$$P_r(s) = \begin{bmatrix} s^2 & 0 \\ 0 & s^2 \end{bmatrix}.$$

step 3. Let us now construct the skeleton column block–matrix, $D^b_R s$, of $D_R(s)$ using the row structure matrix $P_r(s)$. There are $n_i = 2$ active structure rows, $t = n_i.[D_R] - n_s = 2 \times 2 - 4 = 0$ inactive structure rows, which implies that the polynomial matrix $D_R(s)$ is regular, and $n_s = 4$ residue rows. Thus, the skeleton column block–matrix, $D^b_R s$, of $D_R(s)$ is:

$$D^b_R s = \begin{bmatrix} 1 & 0 \\ 0 & 1 \\ - & - \\ e_1 & e_1 \\ e_2 & e_2 \\ - & - \\ e_3 & e_3 \\ e_4 & e_4 \end{bmatrix},$$

where e_i denotes the ith residue row vector of $D^b_R s$.

step 4. Solving

$$\mathcal{AB}_2.D^b_R s = \left[\ A^2.B \quad A.B \quad B\ \right].\begin{bmatrix} D_{R_0} s \\ D_{R_1} s \\ D_{R_2} s \end{bmatrix}$$

$$= \begin{bmatrix} 2 & 27 & | & 2 & -4 & | & 2 & 1 \\ -21 & -18 & | & 7 & -1 & | & 0 & 2 \\ -19 & -50 & | & -1 & 6 & | & 1 & -1 \\ 9 & 20 & | & -7 & -5 & | & -1 & 0 \end{bmatrix} . \begin{bmatrix} 1 & 0 \\ 0 & 1 \\ - & - \\ e_4 & e_4 \\ e_3 & e_3 \\ - & - \\ e_2 & e_2 \\ e_1 & e_1 \end{bmatrix}$$

$$= 0$$

or

$$\begin{bmatrix} 2 & -4 & 2 & 1 \\ 7 & -1 & 0 & 2 \\ -1 & 6 & 1 & -1 \\ -7 & -5 & -1 & 0 \end{bmatrix} \begin{bmatrix} e_4 & e_4 \\ e_3 & e_3 \\ e_2 & e_2 \\ e_1 & e_1 \end{bmatrix} = - \begin{bmatrix} 2 & 27 \\ -21 & -18 \\ -19 & -50 \\ 9 & 20 \end{bmatrix}.$$

Hence

$$D_{\mathrm{R}}^b = \begin{bmatrix} 1.00 & 0.00 \\ 0.00 & 1.00 \\ \hline -3.81 & -7.13 \\ 6.64 & 13.81 \\ \hline 2.51 & 0.83 \\ 27.18 & 40.86 \end{bmatrix}.$$

step 5. The numerator N_{R}^b can now be determined using

$$N_{\mathrm{R}}^b = TL_{\partial[D_{\mathrm{R}}^b]-1}(M_{[D_{\mathrm{R}}^b]-1}^b).\tilde{D}_{\mathrm{R}}^b,$$

where

$$M_1^b = \begin{bmatrix} C.B \\ C.A.B \end{bmatrix} \text{ and } TL_1(M_1^b) = \begin{bmatrix} C.B & 0 \\ C.A.B & C.B \end{bmatrix}.$$

Hence

$$N_{\mathrm{R}}^b = \begin{bmatrix} C.B & 0 \\ C.A.B & C.B \end{bmatrix} \cdot \begin{bmatrix} D_{\mathrm{R}_0} \\ D_{\mathrm{R}_1} \end{bmatrix} = \begin{bmatrix} -1.00 & 2.00 \\ -2.00 & 0.00 \\ 4.00 & 3.00 \\ \hline 34.46 & 38.50 \\ 2.63 & 3.26 \\ 27.65 & 17.92 \end{bmatrix}.$$

The transmission matrix is a zero matrix. Therefore no modification is required to the numerator N_{R}^b.

The block–transpose operation can now be applied to the denominator and numerator column block–matrices D_{R}^b and N_{R}^b to obtain their equivalent row block–matrix descriptions, as shown below.

$$D_{\mathrm{R}} = \left[\begin{array}{cc|cc|cc} 1.00 & 0.00 & -3.81 & -7.13 & 2.51 & 0.83 \\ 0.00 & 1.00 & 6.64 & 13.81 & 27.18 & 40.86 \end{array}\right]$$

and

$$N_{\mathrm{R}} = \left[\begin{array}{cc|cc} -1.00 & 2.00 & 34.46 & 38.50 \\ -2.00 & 0.00 & 2.63 & 3.26 \\ 4.00 & 3.00 & 27.65 & 17.92 \end{array}\right].$$

Thus

$$N_{\mathrm{R}}(s).D_{\mathrm{R}}^{-1}(s) =$$

$$\begin{bmatrix} -s+34.46 & 2s+38.5 \\ -2s+2.63 & 3.26 \\ 4s+27.65 & 3s+17.92 \end{bmatrix} \cdot \begin{bmatrix} s^2-3.81s+2.51 & -7.13s+0.83 \\ 6.64s+27.18 & s^2+13.81s+40.86 \end{bmatrix}^{-1}.$$

To verify the right-coprimeness of the above RMFD, it is sufficient that we show that the eigenvalues of $D_{\mathrm{R}}(s)$ are those of $S(A,B,C,E)$. The eigenvalues of $D_{\mathrm{R}}(s)$ are found to be $\{-0.84 \pm 1.82j,\ -4.16 \pm 1.64j\}$ which are those of $S(A,B,C,E)$. Thus, the above RMFD contains no output-decoupling zeros.

Example 3.2 Consider the following controllable and observable continuous state-space system with $n_s = 4$ states, $n_i = 2$ inputs and $n_o = 4$ outputs.

$$A = \begin{bmatrix} 1 & 0 & 0 & 0 \\ 0 & 2 & 0 & 0 \\ 0 & 0 & 3 & 0 \\ 0 & 0 & 0 & 4 \end{bmatrix}, B = \begin{bmatrix} 1 & 1 \\ 2 & 0 \\ 3 & 0 \\ -6 & 0 \end{bmatrix},$$

$$C = \begin{bmatrix} 1 & 0 & 0 & 0 \\ 0 & 1 & 0 & 0 \\ 0 & 0 & 1 & 0 \\ 0 & 0 & 0 & 1 \end{bmatrix} \text{ and } E = \begin{bmatrix} 0 & -1 \\ 1 & 0 \\ 3 & 2 \\ 0 & 0 \end{bmatrix}.$$

step 1. Using the methodology described in Appendix E the set of controllability indices of the state-space system is $I_r(A,B) = \{\mu_1 = 3,\ \mu_2 = 1\}$.

step 2. The row structure matrix of $D_{\mathrm{R}}(s)$, of dimension 2×2, is

$$P_r(s) = diag\left[\ s^{\mu_1=3},\ \ s^{\mu_2=1}\ \right].I_2$$

or,

$$P_r(s) = \begin{bmatrix} s^3 & 0 \\ 0 & s^1 \end{bmatrix}.$$

step 3. Let us now construct the skeleton column block–matrix, $D^b_{\rm R}s$, of $D_{\rm R}(s)$ using the row structure matrix $P_{\rm r}(s)$. There are $n_{\rm i} = 2$ active structure rows, $t = n_{\rm i}.[D_{\rm R}] - n_{\rm s} = 2 \times 3 - 4 = 2$ inactive structure rows and $n_{\rm s} = 4$ residue rows. Thus, the skeleton column block–matrix, $D^b_{\rm R}s$, of $D_{\rm R}(s)$ is:

$$D^b_{\rm R}s = \begin{bmatrix} 1 & 0 \\ 0 & 0 \\ - & - \\ e_1 & e_1 \\ 0 & 0 \\ - & - \\ e_2 & e_2 \\ 0 & 1 \\ - & - \\ e_3 & e_3 \\ e_4 & e_4 \end{bmatrix},$$

where e_i denotes the ith residue row vector of $D^b_{\rm R}s$.

step 4. Solving

$$\mathcal{AB}_3.D^b_{\rm R}s = \begin{bmatrix} A^3.B & A^2.B & A.B & B \end{bmatrix}.\begin{bmatrix} D_{{\rm R}_0}s \\ D_{{\rm R}_1}s \\ D_{{\rm R}_2}s \\ D_{{\rm R}_3}s \end{bmatrix}$$

$$= \left[\begin{array}{cc|cc|cc|cc} 1 & 1 & 1 & 1 & 1 & 1 & 1 & 1 \\ 16 & 0 & 8 & 0 & 4 & 0 & 2 & 0 \\ 81 & 0 & 27 & 0 & 9 & 0 & 3 & 0 \\ -384 & 0 & -96 & 0 & -24 & 0 & -6 & 0 \end{array}\right].\begin{bmatrix} 1 & 0 \\ 0 & 0 \\ - & - \\ e_1 & e_1 \\ 0 & 0 \\ - & - \\ e_2 & e_2 \\ 0 & 1 \\ - & - \\ e_3 & e_3 \\ e_4 & e_4 \end{bmatrix}$$

$$= 0$$

or, equivalently,

$$\begin{bmatrix} 1 & 1 & 1 & 1 \\ 8 & 4 & 2 & 0 \\ 27 & 9 & 3 & 0 \\ -96 & -24 & -6 & 0 \end{bmatrix}.\begin{bmatrix} e_1 & e_1 \\ e_2 & e_2 \\ e_3 & e_3 \\ e_4 & e_4 \end{bmatrix} = -\begin{bmatrix} 1 & 1 \\ 16 & 4 \\ 81 & 9 \\ -384 & -24 \end{bmatrix}.$$

Hence we find

$$D_{\mathrm{R}}^{b} = \begin{bmatrix} 1.00 & 0.00 \\ 0.00 & 0.00 \\ ---- & ---- \\ -9.00 & 0.00 \\ 0.00 & 0.00 \\ ---- & ---- \\ 26.00 & 0.00 \\ 0.00 & 1.00 \\ ---- & ---- \\ -24.00 & 0.00 \\ 6.00 & -1.00 \end{bmatrix}.$$

step 5. The numerator N_{R}^{b} can now can be determined using

$$N_{\mathrm{R}}^{b} = TL_{\partial[D_{\mathrm{R}}^{b}]-1}(M_{[D_{\mathrm{R}}^{b}]-1}^{b}).\tilde{D}_{\mathrm{R}}^{b},$$

where

$$M_2^b = \begin{bmatrix} C.B \\ C.A.B \\ C.A^2.B \end{bmatrix} \text{ and } TL_2(M_2^b) = \begin{bmatrix} C.B & 0 & 0 \\ C.A.B & C.B & 0 \\ C.A^2.B & C.A.B & C.B \end{bmatrix}.$$

Thus

$$N_{\mathrm{R}}^{b} = \begin{bmatrix} C.B & 0 & 0 \\ C.A.B & C.B & 0 \\ C.A^2.B & C.A.B & C.B \end{bmatrix} . \begin{bmatrix} D_{\mathrm{R}_0} \\ D_{\mathrm{R}_1} \\ D_{\mathrm{R}_2} \end{bmatrix}$$

$$= \begin{bmatrix} 1.00 & 0.00 \\ 2.00 & 0.00 \\ 3.00 & 0.00 \\ -6.00 & 0.00 \\ ---- & ---- \\ -8.00 & 0.00 \\ -14.00 & 0.00 \\ -18.00 & 0.00 \\ 30.00 & 0.00 \\ ---- & ---- \\ 18.00 & 1.00 \\ 24.00 & 0.00 \\ 24.00 & 0.00 \\ -36.00 & 0.00 \end{bmatrix}.$$

step 6. The transmission matrix, E, of the state–space system is non–zero. Therefore, the numerator column block–matrix is modified to

$$\begin{bmatrix} 0_{n_o \times n_i} \\ N_{\mathrm{R}}^{b} \end{bmatrix} + < E >_{\partial[D_{\mathrm{R}}]}^{b} .D_{\mathrm{R}}^{b} .$$

Hence

$$N_{\mathrm{R}}^{b} = \left[\begin{array}{cc} 0.00 & 0.00 \\ 0.00 & 0.00 \\ 0.00 & 0.00 \\ 0.00 & 0.00 \\ \hline 1.00 & 0.00 \\ 2.00 & 0.00 \\ 3.00 & 0.00 \\ -6.00 & 0.00 \\ \hline -8.00 & 0.00 \\ -14.00 & 0.00 \\ -18.00 & 0.00 \\ 30.00 & 0.00 \\ \hline 18.00 & 1.00 \\ 24.00 & 0.00 \\ 24.00 & 0.00 \\ -36.00 & 0.00 \end{array}\right] + \left[\begin{array}{cc} 0.00 & 0.00 \\ 1.00 & 0.00 \\ 3.00 & 0.00 \\ 0.00 & 0.00 \\ \hline 0.00 & 0.00 \\ -9.00 & 0.00 \\ -27.00 & 0.00 \\ 0.00 & 0.00 \\ \hline 0.00 & -1.00 \\ 26.00 & 0.00 \\ 78.00 & 2.00 \\ 0.00 & 0.00 \\ \hline -6.00 & 1.00 \\ -24.00 & 0.00 \\ -60.00 & -2.00 \\ 0.00 & 0.00 \end{array}\right]$$

$$= \left[\begin{array}{cc} 0.00 & 0.00 \\ 1.00 & 0.00 \\ 3.00 & 0.00 \\ 0.00 & 0.00 \\ \hline 1.00 & 0.00 \\ -7.00 & 0.00 \\ -24.00 & 0.00 \\ -6.00 & 0.00 \\ \hline -8.00 & -1.00 \\ 12.00 & 0.00 \\ 60.00 & 2.00 \\ 30.00 & 0.00 \\ \hline 12.00 & 2.00 \\ 0.00 & 0.00 \\ -36.00 & -2.00 \\ -36.00 & 0.00 \end{array}\right]$$

Using the block–transpose operation on the denominator and numerator column block-matrices, D_{R}^{b} and N_{R}^{b}, we realise their equivalent row block–matrix descriptions, as shown below.

$$D_{\mathrm{R}} = \begin{bmatrix} 1.00 & 0.00 & | & -9.00 & 0.00 & | & 26.00 & 0.00 & | & -24.00 & 0.00 \\ 0.00 & 0.00 & | & 0.00 & 0.00 & | & 0.00 & 1.00 & | & 6.00 & -1.00 \end{bmatrix}$$

and

$$N_{\mathrm{R}} = \begin{bmatrix} 0.00 & 0.00 & | & 1.00 & 0.00 & | & -8.00 & -1.00 & | & 12.00 & 2.00 \\ 1.00 & 0.00 & | & -7.00 & 0.00 & | & 12.00 & 0.00 & | & 0.00 & 0.00 \\ 3.00 & 0.00 & | & -24.00 & 0.00 & | & 60.00 & 2.00 & | & -36.00 & -2.00 \\ 0.00 & 0.00 & | & -6.00 & 0.00 & | & 30.00 & 0.00 & | & -36.00 & 0.00 \end{bmatrix}.$$

Thus

$$N_{\mathrm{R}}(s).D_{\mathrm{R}}^{-1}(s) =$$

$$\begin{bmatrix} s^2 - 8s + 12 & -s + 2 \\ s^3 - 7s^2 + 12s & 0 \\ 3s^3 - 24s^2 + 60s - 36 & 2s - 2 \\ -6s^2 + 30s - 36 & 0 \end{bmatrix} . \begin{bmatrix} s^3 - 9s^2 + 26s - 24 & 0 \\ 6 & s - 1 \end{bmatrix}^{-1}.$$

To verify the right–coprimeness of the above RMFD, it is sufficient that we show that the eigenvalues of $D_{\mathrm{R}}(s)$ are those of $S(A, B, C, E)$. The eigenvalues of $D_{\mathrm{R}}(s)$ are found to be $\{1,\ 2,\ 3,\ 4\}$ which are those of $S(A, B, C, E)$. Thus, the above RMFD contains no output–decoupling zeros.

3.3.3 Summary

The above methodology is simple because the only operations involved are scalar matrix addition and multiplication, and the solving of a set of linear equations. The method operates on the matrix of coefficients of the MFD and SSD rather than on their polynomial matrices. Hence the calculations with polynomials are replaced by operations on real matrices which normally have better numerical properties. Note that $D_{\mathrm{R}}(q)$ and $N_{\mathrm{R}}(q)$ are not unique.

3.4 The Inverse of a Polynomial Matrix

3.4.1 Introduction

The inverse of a polynomial matrix is required in section 3.5 in which a MFD is transformed to a transfer function matrix description (TFMD).

The algorithm proposed for computing the inverse of a polynomial matrix $L(q)$ employs the techniques of linear transformations, the method of resultants and the pseudo–inverse of a resultant.

Proposition 3.3 Consider a $n \times n$ polynomial matrix $L(q)$ of rank n and degree m, the determinant of which is $l(q) = det(L(q))$ of polynomial order $\partial[l(q)]$.

The inverse of the polynomial matrix $L(q)$ is defined by

$$L^{-1}(q) = \frac{\mathrm{Adj}(L(q))}{l(q)} . \tag{3.32}$$

From eqn. (3.32) the inverse of $L(q)$ is composed of two functions, the adjoint and the determinant of $L(q)$. The algorithm developed in § 2.4.8 enables us to compute the determinant of $L(q)$ given its row–block matrix form L. An algorithm which calculates the adjoint of $L(q)$, given the row block–matrix description of $L(q)$, will now be described. By combining the two algorithms the inverse of $L(q)$ is realised.

3.4.2 Computing the Adjoint of a Polynomial Matrix

The numerical method proposed for the calculation of the adjoint, $\mathrm{Adj}(L(q))$, of a polynomial matrix, $L(q)$, utilises a linear systems transformation and the fact that the adjoint of a polynomial matrix is also a polynomial matrix.

The square polynomial matrix $L(q)$ can be thought of as a denominator of a transfer function of a linear system. The determinant polynomial $l(q) = det(L(q))$, the zeros of which correspond to the eigenvalues of $L(q)$, is the characteristic polynomial of such a system. The characteristic polynomial $l(q)$ is unique and independent of the linear system description.

Let the inverse of the polynomial matrix $L(q)$ be modelled as a n-input n-output RMFD; i.e.,

$$G_{\mathrm{R}}(q) = I_n.L^{-1}(q) . \tag{3.33}$$

The above RMFD, $G_{\mathrm{R}}(q)$, can be transformed to its equivalent LMFD

$$G_{\mathrm{L}}(q) = (l(q) \otimes I_n)^{-1}.Ad(q), \tag{3.34}$$

where I_n is the $n \times n$ identity matrix and $Ad(q)$ is a polynomial matrix denoting the adjoint of $L(q)$, $\mathrm{Adj}(L(q))$, the degree and dimension of which are $\partial[Ad]$ and $n \times n$, respectively. Let us formulate the problem in terms of exact model matching with respect to the transfer functions given by eqns. (3.33) and (3.34); i.e.,

$$\left[\begin{array}{cc} -(l(q) \otimes I_n) & Ad(q) \end{array}\right] \cdot \left[\begin{array}{c} I_n \\ L(q) \end{array}\right] = 0 \,. \tag{3.35}$$

By application of the method of resultants to eqn. (3.35), the equivalent numerical description is

$$\left[\begin{array}{cc} -(l \otimes I_n) & Ad \end{array}\right] \cdot \left[\begin{array}{c} [0_{n.(\partial[Ad]+m-\partial[l])} \;\; < I_n >_{\partial[l]}] \\ < L >_{\partial[Ad]} \end{array}\right] = 0, \tag{3.36}$$

where l and Ad denote the row block–matrix descriptions of $l(q)$ and $Ad(q)$, respectively. The dimensions of the two resultants are as follows:

$< I_n >_{\partial[l]} = I_{n.(\partial[l]+1)}$ is of dimension $n.(\partial[l]+1) \times n.(\partial[l]+1)$; and

$< L >_{\partial[Ad]}$ is of dimension $n.(\partial[Ad]+1) \times n.(\partial[Ad]+m+1)$.

The $n.(\partial[Ad]+m-\partial[l])$-column matrix of zeros $0_{n.(\partial[Ad]+m-\partial[l])}$ is required in eqn. (3.36) because, in general, the column dimension of the resultant $< I_n >_{\partial[l]}$ is less than the column dimension of the resultant $< L >_{\partial[Ad]}$.

The exact degree of the adjoint polynomial matrix $Ad(q)$ is unknown, therefore we must assume the maximum possible degree for the adjoint polynomial matrix of a $n \times n$ polynomial matrix of degree $\partial[L] = m$; this is $\partial[Ad] = m.(n-1)$. Using the maximum possible degree for $Ad(q)$ some of the most significant coefficient block–matrices of the solution to the row block–matrix Ad may be zero matrices.

Equation (3.36) can be simplified by partial separation of the row block–matrices l and Ad. The technique is based on the properties of the pseudo–inverses of a resultant. Using the methodology described by Appendix D, the $n.(\partial[Ad]+m+1) \times n.(\partial[Ad]+m+1)$ pseudo–inverse $Z_0^{-1}(< L >_{\partial[Ad]})$, of order $k = 0$, of the resultant $< L >_{\partial[Ad]}$ is

$$Z_0^{-1}(< L >_{\partial[Ad]}) = \left[\begin{array}{c} < L >_{\partial[Ad]} \\ B \end{array}\right]^{-1} , \tag{3.37}$$

where B is a $n.m \times n.(\partial[Ad] + m + 1)$ matrix chosen such that $Z_0(< L >_{\partial[Ad]})$ is non–singular, and

$$< L >_{\partial[Ad]} .Z_0^{-1}(< L >_{\partial[Ad]}) = \begin{bmatrix} I_{n.(\partial[Ad]+1)} & 0_{n.m} \end{bmatrix} . \tag{3.38}$$

Let the pseudo–inverse, $Z_0^{-1}(< L >_{\partial[Ad]})$, take on the form

$$Z_0^{-1}(< L >_{\partial[Ad]}) = \begin{bmatrix} X_{11} & | & X_{12} \\ - & + & - \\ X_{21} & | & X_{22} \end{bmatrix},$$

where X_{21} and X_{22} are of dimension $n.(\partial[l] + 1) \times n.(\partial[Ad] + 1)$ and $n.(\partial[l] + 1) \times n.m$, respectively.

Postmultiplying eqn. (3.36) by the pseudo–inverse, $Z_0^{-1}(< L >_{\partial[Ad]})$, of eqn. (3.37) we obtain

$$\begin{bmatrix} -(l \otimes I_n) & Ad \end{bmatrix} . \begin{bmatrix} X_{21} & X_{22} \\ I_{n.(\partial[Ad]+1)} & 0_{n.m} \end{bmatrix} = 0, \tag{3.39}$$

or, equivalently,

$$Ad = (l \otimes I_n).X_{21} \tag{3.40}$$

and

$$(l \otimes I_n).X_{22} = 0 . \tag{3.41}$$

As previously indicated the determinant polynomial $l(q)$ is equivalent to the characteristic polynomials of the linear systems eqns. (3.33) and (3.34). Applying the methodology of § 2.4.8, the determinant polynomial $l(q) = det(L(q))$ can be computed. Then eqn. (3.40) can be used to calculate the adjoint polynomial matrix $Ad(q)$ of $L(q)$.

Algorithm 3.2 The above methodology, which computes the inverse of a $n \times n$ non–singular polynomial matrix $L(q)$, can be summarised as follows:

step 1. Set $\partial[Ad] = \partial[L].(n - 1)$.

step 2. Construct the matrix $Z_0(< L >_{\partial[Ad]})$, for $k = 0$, of the resultant $< L >_{\partial[Ad]}$ (refer to Appendix D). Take the inverse of the matrix $Z_0(< L >_{\partial[Ad]})$ using a routine from a software package or mathematics library.

step 3. Compute the polynomial $l(q) = det(L(q))$ using Algorithm 2.3 of § 2.4.8.

step 4. Extract the $n.(\partial[l]+1) \times n.(\partial[Ad]+1)$ matrix X_{21} from $Z_0^{-1}(< L >_{\partial[Ad]})$.

step 5. Calculate $Ad = (l \otimes I_n).X_{21}$.

Example 3.3 Consider a polynomial matrix $L(q)$ of degree $\partial[L(q)] = 2$, the row block–matrix, L, of which is shown below. The inverse of the polynomial matrix $L(q)$ is to be determined using the algorithm described above. The computations are carried out with the aid of a software package [53] using double precision arithmetic.

Let

$$L = \left[\begin{array}{cc|cc|cc} 1.00 & 0.00 & -3.81 & -7.13 & 2.51 & 0.83 \\ 0.00 & 1.00 & 6.64 & 13.81 & 27.18 & 40.86 \end{array}\right].$$

step 1. The maximum degree of the adjoint polynomial matrix, $Ad(q)$, of $L(q)$ is $\partial[Ad] = m.(n-1) = 2$.

step 2. Using the methodology described in Appendix D the pseudo–inverse, for $k = 0$, of the 6×10 resultant $< L >_2$ is

$$Z_0^{-1}(< L >_2) = \left[\begin{array}{ccccccc} 1.0 & 0.0 & 3.81 & 7.13 & -35.34 & -72.13 & 140.95 \\ 0.0 & 1.0 & -6.64 & -13.81 & 39.22 & 102.51 & -139.24 \\ 0.0 & 0.0 & 1.0 & 0.0 & 3.81 & 7.13 & -35.34 \\ 0.0 & 0.0 & 0.0 & 1.0 & -6.64 & -13.81 & 39.22 \\ 0.0 & 0.0 & 0.0 & 0.0 & 1.0 & 0.0 & 3.81 \\ 0.0 & 0.0 & 0.0 & 0.0 & 0.0 & 1.0 & -6.64 \\ 0.0 & 0.0 & 0.0 & 0.0 & 0.0 & 0.0 & 1.0 \\ 0.0 & 0.0 & 0.0 & 0.0 & 0.0 & 0.0 & 0.0 \\ 0.0 & 0.0 & 0.0 & 0.0 & 0.0 & 0.0 & 0.0 \\ 0.0 & 0.0 & 0.0 & 0.0 & 0.0 & 0.0 & 0.0 \end{array}\right. \cdots$$

$$\cdots \left.\begin{array}{ccc} 449.67 & 2049.19 & 2976.56 \\ -566.38 & -2884.74 & -4221.23 \\ -72.13 & -203.36 & -294.49 \\ 102.51 & 392.02 & 569.79 \\ 7.13 & -2.51 & -0.83 \\ -13.81 & -27.18 & -40.86 \\ 0.0 & 0.0 & 0.0 \\ 1.0 & 0.0 & 0.0 \\ 0.0 & 1.0 & 0.0 \\ 0.0 & 0.0 & 1.0 \end{array}\right].$$

step 3. Applying Algorithm 2.3 of § 2.4.8, the determinant polynomial, $l(q)$, of $L(q)$ is

$$l(q) = q^4 + 10q^3 + 38.10q^2 + 67.27q + 80,$$

the polynomial order of which is 4.

step 4. Extracting the 10×6 matrix X_{21} from the pseudo-inverse $Z_0^{-1}(< L >_2)$, we have

$$X_{21} = \begin{bmatrix} 1.0 & 0.0 & 3.81 & 7.13 & -35.34 & -72.13 \\ 0.0 & 1.0 & -6.64 & -13.81 & 39.22 & 102.51 \\ 0.0 & 0.0 & 1.0 & 0.0 & 3.81 & 7.13 \\ 0.0 & 0.0 & 0.0 & 1.0 & -6.64 & -13.81 \\ 0.0 & 0.0 & 0.0 & 0.0 & 1.0 & 0.0 \\ 0.0 & 0.0 & 0.0 & 0.0 & 0.0 & 1.0 \\ 0.0 & 0.0 & 0.0 & 0.0 & 0.0 & 0.0 \\ 0.0 & 0.0 & 0.0 & 0.0 & 0.0 & 0.0 \\ 0.0 & 0.0 & 0.0 & 0.0 & 0.0 & 0.0 \\ 0.0 & 0.0 & 0.0 & 0.0 & 0.0 & 0.0 \end{bmatrix}.$$

step 5. Calculate $Ad = (l \otimes I_2).X_{21}$; i.e.,

$$Ad = \begin{bmatrix} 1.00 & 0.00 & 10.00 & 0.00 & 38.10 & 0.00 & 67.27 & 0.00 & 80.00 & 0.00 \\ 0.00 & 1.00 & 0.00 & 10.00 & 0.00 & 38.10 & 0.00 & 67.27 & 0.00 & 80.00 \end{bmatrix} .X_{21}$$

$$= \left[\begin{array}{cc|cc|cc} 1.00 & 0.00 & 13.81 & 7.13 & 40.86 & -0.83 \\ 0.00 & 1.00 & -6.64 & -3.81 & -27.18 & 2.51 \end{array}\right]$$

Hence the inverse of $L(q)$ is

$$L^{-1}(q) = Ad(q)/l(q)$$

$$= \frac{\begin{bmatrix} q^2 + 13.81q + 40.86 & 7.13q - 0.83 \\ -6.64q - 27.18 & q^2 - 3.81q + 2.51 \end{bmatrix}}{q^4 + 10q^3 + 38.10q^2 + 67.27q + 80}.$$

3.4.3 Summary

The above methodology is simple because the operations involve only addition and multiplication of real matrices. The algorithm is well-conditioned due to the use, in step 2, of a numerically robust routine for forming the inverse of a real matrix [30].

3.5 Transforming a MFD to TFMD

3.5.1 Introduction

The transfer function matrix description (TFMD) of a plant or system model is of the form

$$G(q) = \frac{N(q)}{d(q)},$$

where $N(q) \in \Re[q]^{n_o \times n_i}$ is the numerator polynomial matrix and $d(q) \in \Re[q]$ is a scalar polynomial. Furthermore, if $G(q)$ is coprime we can identify $d(q)$ as the characteristic polynomial of $G(q)$. Then the zeros of $d(q)$ are the poles of $G(q)$, and the polynomial order of $d(q)$ is equal to the McMillan degree of $G(q)$. In order to satisfy the conditions of causality, the degree of the numerator polynomial matrix, $N(q)$, must be less than or equal to the degree of the scalar polynomial $d(q)$; i.e., $\partial[N(q)] \leq \partial[d(q)]$.

3.5.2 An Algorithm to transform a MFD to TFMD

The problem of transforming a MFD to its equivalent TFMD is formulated in terms of exact model matching; i.e.,

$$G(q) = \frac{N(q)}{d(q)} = N_{\mathrm{R}}(q).D_{\mathrm{R}}^{-1}(q) = D_{\mathrm{L}}^{-1}(q).N_{\mathrm{L}}(q),$$

or

$$G(q) = \frac{N(q)}{d(q)} = \frac{N_{\mathrm{R}}(q).\mathrm{Adj}(D_{\mathrm{R}}(q))}{det(D_{\mathrm{R}}(q)} = \frac{\mathrm{Adj}(D_{\mathrm{L}}(q)).N_{\mathrm{L}}(q)}{det(D_{\mathrm{L}}(q))},$$

where $G(q)$ is unique to within a least common multiple.

Algorithm 3.3 The algorithm proposed in transforming a MFD to its equivalent TFMD is described by the following two steps:

step 1. Compute the inverse of the denominator polynomial matrix of the MFD, (see § 3.4).

step 2. Calculate the numerator polynomial matrix, $N(q)$, of the TFMD by computing the product of the numerator polynomial matrix of the MFD and the adjoint of the denominator polynomial matrix of the MFD (see § 2.4.5).

$\nabla\nabla\nabla$

The following example implements the above algorithm to transform a RMFD to its equivalent TFMD. The computations are carried out with aid of a software package [53] using double precision arithmetic.

Example 3.4 Consider the following row block–matrix description of a right–coprime RMFD, which is to be transformed to its equivalent TFMD.

$$D_{\mathrm{R}} = \left[\begin{array}{cc|cc|cc} 1.00 & 0.00 & -3.81 & -7.13 & 2.51 & 0.83 \\ 0.00 & 1.00 & 6.64 & 13.81 & 27.18 & 40.86 \end{array}\right] \text{ and}$$

$$N_{\mathrm{R}} = \left[\begin{array}{cc|cc} 1.00 & 2.00 & 34.46 & 38.50 \\ -2.00 & 0.00 & 2.63 & 3.26 \\ 4.00 & 3.00 & 27.65 & 17.92 \end{array}\right] .$$

The numerator, $N_{\mathrm{R}}(q)$, and denominator, $D_{\mathrm{R}}(q)$, are 3×2 and 2×2 polynomial matrices of degree $\partial[N_{\mathrm{R}}] = 1$ and $\partial[D_{\mathrm{R}}] = 2$, respectively.

step 1. Compute the inverse of $D_{\mathrm{R}}(q)$ by applying Algorithm 3.2 of § 3.4.

The polynomial matrix $D_{\mathrm{R}}(q)$ is monic. Therefore, we can directly form its companion matrix; i.e.,

$$C_m(D_{\mathrm{R}}(q)) = \begin{bmatrix} 3.81 & 7.13 & -2.51 & -0.83 \\ -6.64 & -13.81 & -27.18 & -40.86 \\ 1 & 0 & 0 & 0 \\ 0 & 1 & 0 & 0 \end{bmatrix} .$$

The characteristic polynomial of $q.I_4 - C_m(D_{\mathrm{R}}(q))$ is

$$d(q) = q^4 + 10q^3 + 38.10q^2 + 67.27q + 80 .$$

Compute the adjoint $Ad(q)$ of $D_{\mathrm{R}}(q)$. The row block–matrix description of the adjoint of $D_{\mathrm{R}}(q)$ is found to be

$$Ad = \left[\begin{array}{cc|cc|cc} 1.00 & 0.00 & 13.81 & 7.13 & 40.86 & -0.83 \\ 0.00 & 1.00 & -6.64 & -3.81 & -27.18 & 2.51 \end{array}\right] .$$

step 2. The numerator row block–matrix description N is calculated as follows:

$$N = N_{\mathrm{R}}. < Ad >_{\partial[N_{\mathrm{R}}]}$$

$$= \left[\begin{array}{cc|cc} 1.00 & 2.00 & 34.46 & 38.50 \\ -2.00 & 0.00 & 2.63 & 3.26 \\ 4.00 & 3.00 & 27.65 & 17.92 \end{array}\right] . \left[\begin{array}{cc|cc|c} 1.00 & 0.00 & 13.81 & 7.13 & \\ 0.00 & 1.00 & -6.64 & -3.81 & \cdots \\ 0.00 & 0.00 & 1.00 & 0.00 & \\ 0.00 & 0.00 & 0.00 & 1.00 & \end{array}\right.$$

$$\cdots\begin{array}{cc|cc} 40.86 & -0.83 & 0.00 & 0.00 \\ -27.18 & 2.51 & 0.00 & 0.00 \\ 13.81 & 7.13 & 40.86 & -0.83 \\ -6.64 & -3.81 & -27.18 & 2.51 \end{array}\Bigg]$$

$$= \left[\begin{array}{cc|cc|cc|cc} 1.00 & 2.00 & 34.99 & 38.01 & 206.75 & 103.21 & 361.61 & 68.03 \\ -2.00 & 0.00 & -24.99 & -11.00 & -67.05 & 7.99 & 18.86 & 6.00 \\ 4.00 & 3.00 & 62.97 & 35.01 & 344.76 & 133.08 & 642.71 & 22.03 \end{array}\right].$$

Hence the equivalent TFMD of the RMFD is

$$G(q) = \frac{\left[\begin{array}{cc} q^3+34.99q^2+206.75q+361.61 & 2q^3+38.01q^2+103.21q+68.03 \\ -2q^3-24.99q^2-67.05q+18.86 & -11q^2+7.99q+6.00 \\ 4q^3+62.97q^2+344.76q+642.71 & 3q^3+35.01q^2+133.08q+22.03 \end{array}\right]}{q^4+10q^3+38.10q^2+67.27q+80.00}.$$

3.5.3 Summary

The method operates on the matrix of coefficients of the numerator and denominator of the MFDs rather than on the polynomial matrices. Hence the calculations with polynomials are replaced by calculations with real matrices which normally yield better numerical properties.

3.6 The AutoRegressive–Moving Average Model

3.6.1 Introduction

In Chapter 4, a method for the design of a dynamic compensator for a MIMO system is presented. For a discrete system, the AutoRegressive–Moving Average (ARMA) model is used in the construction of the control algorithm of a digital compensator. The compensator, however, is described by its LMFD and transformation to the ARMA model is thus required. The output of a digital compensator can be represented as a function of past outputs and past and present inputs. In general, an ARMA model is usually more easily implemented by a recursive realisation using feedback.

3.6.2 Canonical Monic Form of ARMA Model

In control engineering, it is well known that a MFD of a system is closely connected to the ARMA model of the same system. This has been discussed in a paper by Gevers [48]. Gevers, in the same paper, also investigates a number of limitations when deriving ARMA models from coprime MFD models. He shows that monic ARMA models (see eqn. (3.44)) can only represent systems the McMillan degree of which is a integer multiple of the number of inputs (outputs) for a RMFD (LMFD) model. This result is substantiated by Result 3.1 (Result 3.3).

Assume the transfer function form of interest to be the LMFD,

$$c(t) = D_{\mathrm{L}}^{-1}(z).N_{\mathrm{L}}(z).u(t) = [D_{\mathrm{L}_0} z^m + \sum_{i=1}^{m} D_{\mathrm{L}_i} z^{m-i}]^{-1}.\sum_{i=0}^{m} N_{\mathrm{L}_i} z^{m-i}.u(t), \tag{3.42}$$

where D_{L_0} is non-singular (i.e., $D_{\mathrm{L}}(q)$ is regular) and $det(D_{\mathrm{L}}(q)) \not\equiv 0$. Multiply the numerator and denominator of eqn. (3.42) by z^{-m}; the difference equation corresponding to eqn. (3.42) is then given by

$$c(t) = -\sum_{i=1}^{m} \overline{D}_{\mathrm{L}_i} z^{-i}.c(t) + \sum_{i=0}^{m} \overline{N}_{\mathrm{L}_i} z^{-i}.u(t) \tag{3.43}$$

or, equivalently,

$$c(t) = -\sum_{i=1}^{m} \overline{D}_{\mathrm{L}_i}.c(t-i) + \sum_{i=0}^{m} \overline{N}_{\mathrm{L}_i}.u(t-i), \tag{3.44}$$

where $\overline{D}_{\mathrm{L}_i} = D_{\mathrm{L}_0}^{-1}.D_{\mathrm{L}_i}$ and $\overline{N}_{\mathrm{L}_i} = D_{\mathrm{L}_0}^{-1}.N_{\mathrm{L}_i}$. Such models will be called monic ARMA models.

With monic ARMA models it is very easy to construct one-step ahead predictors. Hence monic ARMA models are widely used with prediction error identification methods, parameter estimation and adaptive control [3,56,10].

The LMFD in the above calculations is assumed to be regular, i.e., the leading coefficient of $D_{\mathrm{L}}(z)$ is non-singular, but this is not in general the case. If, in turn, the LMFD is given by a pair $\widehat{N}_{\mathrm{L}}(z)$ and $\widehat{D}_{\mathrm{L}}(z)$ such that $\widehat{D}_{\mathrm{L}}(z)$ is not regular, then the denominator matrix can be transformed into a regular polynomial matrix by pre-multiplication of the numerator and denominator by a nilpotent matrix $T(z)$, such that $N_{\mathrm{L}}(z) = T(z).\widehat{N}_{\mathrm{L}}(z)$

and $D_L(z) = T(z).\widehat{D}_L(z)$ [90] (see Appendix C). Then, the realisation of the ARMA model of the equivalent LMFD

$$(T(z).\widehat{D}_L(z))^{-1}.(T(z).N_L(z)) = D_L^{-1}(z).N_L(z)$$

can proceed.

Of the number of well known equivalent configurations of the ARMA model, probably the most common is the canonical form shown in Fig. 3.1. This realisation represents a direct implementation of eqn. (3.44). Notice that the delay operations on the left provide successively shifted values of $u(t)$, while those on the right provide the comparable values of $c(t)$. One disadvantage of the direct form method is that it may require up to $2m$ delay elements or operations for a mth order system.

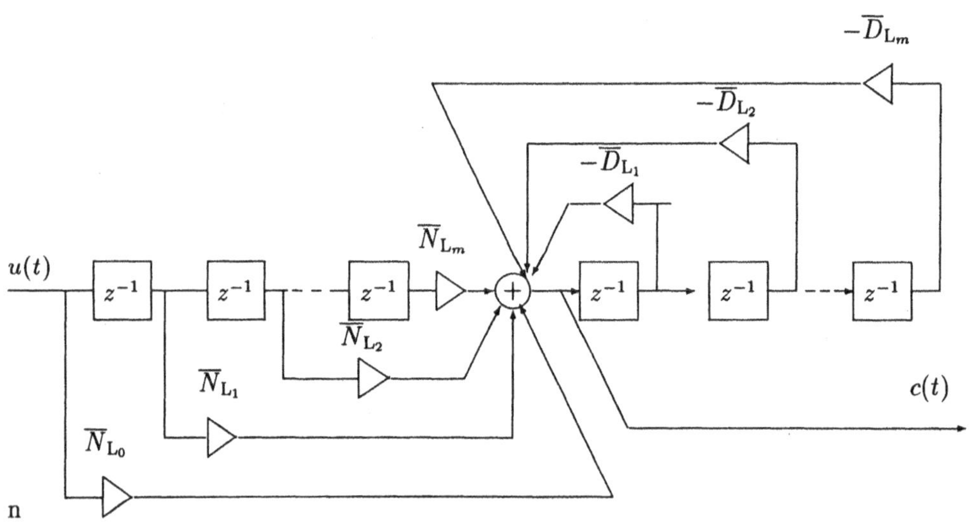

Figure 3.1: Layout of the direct form realisation

Comment: A procedure similar to the one used on the LMFD can be applied to the RMFD to realise its equivalent ARMA model and canonic form realisation. Furthermore, the canonic form of the RMFD, the difference equation of which is of order m, requires up to $m+1$ delay elements or operations, in comparison to $2m$ delay elements or operations for the canonic form of the LMFD.

Chapter 4

The Design of the Closed–Loop System Compensator

4.1 Introduction

In this chapter we will present a computational approach for the design of a two–element compensator [114], which is shown in Figure 1.1 and consists of a dynamic output–feedback compensator, $F(q)$, and a pre–compensator, $R(q)$. The design of the feedback compensator employs pole assignment to determine the closed–loop modes; the parameters of the pre–compensator are derived from the steady–state performance specifications.

In the implementation of the compensator the designer is required to satisfy certain objectives for the performance of the closed–loop system. A set of such objectives may include

- stability;
- speed, overshoot, damping, settling time of the response;
- steady–state accuracy, static or tracking;
- insensitivity to plant or compensator parameter variations;
- insensitivity to external disturbances;
- decoupling; and
- integrity.

The design of control systems by pole assignment has two main features – the choice of pole locations to meet the design specifications and the design of the compensator to realise the desired pole locations. When employing pole assignment it should be remembered that it determines only the modes in the closed–loop system response. Since this is but one aspect of the design, pole assignment is a somewhat restricted, specialised design technique.

It is well–established that for single–input systems, pole assignment can be used to specify the closed–loop transient response completely [81]. This is because the zeros are invariant under feedback. Also the compensator is unique for a given set of closed–loop poles. In MIMO systems the situation is very different; the zeros are moved in a complicated way by feedback, and the compensator is not unique.

The main deficiencies in pole–assignment techniques are the difficulty in assigning the zeros in MIMO systems and the lack of any methodology for selecting the best compensator out of an infinite set which, at the same time, satisfies the pole–assignment and other design requirements. Some control over the closed–loop zero locations can be provided through the design of the pre–compensator.

Pole placement by output feedback in a linear time–invariant multivariable system has attracted considerable attention in recent years (Kimura [61,62], Davison [22], Davison and Wang [23], Brasch and Pearson [14], Pearson and Ding [95], Chen and Hsu [19], Chen [18], Kailath [56], Munro [81]) and Papliński and Gibbard [89]). The field of dynamic output–feedback control is an established discipline, however, the gap between theoretical considerations and engineering, or at least computational techniques, is still wide even in the case of static output feedback. Pending a complete solution of the zeros problem, one procedure, which is a practical solution, is to use pole assignment to design a feedback compensator to provide the desired closed–loop modes and to employ a pre–compensator to satisfy the steady–state requirements. The resulting transient and steady–state performance of the closed–loop system can be assessed with the aid of computer graphics and, if necessary, the design modified. The design process thus requires fast algorithms for efficient interaction.

The design and synthesis of the dynamic output–feedback compensator employs the matrix fraction representation and numerical operations on polynomial matrices to realise a least–order, proper (causal), coprime output–feedback compensator. The computational methods used in the design process are based on the algebraic properties of the system, thus the results are valid for both continuous– and discrete–time systems.

4.2 The Closed–Loop System

4.2.1 The System Configuration

The configuration proposed for the closed-loop system (see Fig. 4.1) consists of the plant, the two–element compensator and a feedback interconnection matrix.

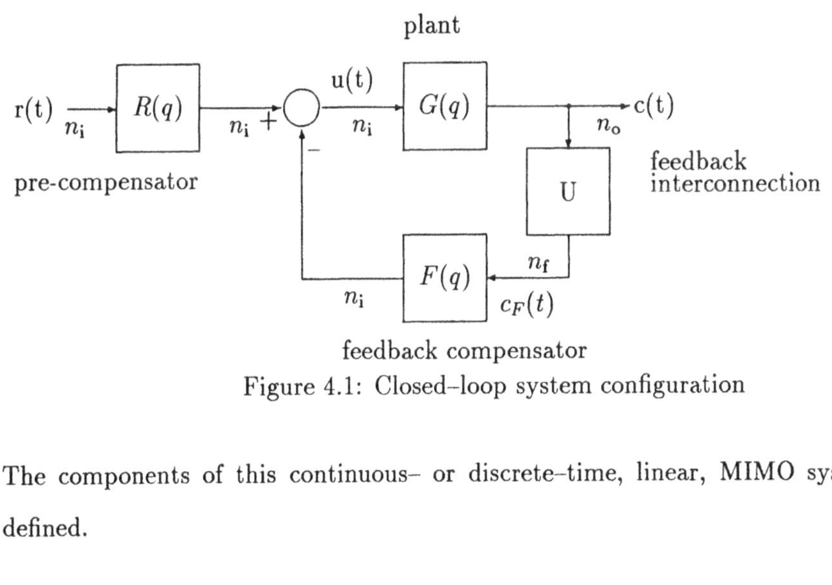

Figure 4.1: Closed–loop system configuration

The components of this continuous– or discrete–time, linear, MIMO system are now defined.

$G(q) \in \Re(q)^{n_o \times n_i}$ is a strictly proper (strictly causal), least–order plant which has n_i actuating inputs and the n_o controlled outputs. The design methodology requires $G(q)$ be expressed as a right–coprime RMFD; i.e.,

$$G(q) = N_R(q).D_R^{-1}(q)$$

where $D_R(q) \in \Re[q]^{n_i \times n_i}$ is the non-singular denominator and $N_R(q) \in \Re[q]^{n_o \times n_i}$ the numerator. The characteristic polynomial, $d(q)$, of $G(q)$ is defined as the determinant of $D_R(q)$; i.e.,

$$d(q) \triangleq \det(D_R(q))\,.$$

The polynomial order, $n_s = \partial[d(q)]$, is the McMillan degree of $G(q)$. Further, let $S(A, B, C)$ denote the SSD of $G(q)$.

$F(q) \in \Re(q)^{n_f \times n_i}$ is a proper (causal), least-order output-feedback compensator, the inputs to which are n_f feedback signals from the plant, $G(q)$. The design methodology requires $F(q)$ be described by a left-coprime LMFD; i.e.,

$$F(q) = V^{-1}(q).K(q)$$

where $V(q) \in \Re[q]^{n_i \times n_i}$ is the non-singular denominator and $K(q) \in \Re[q]^{n_i \times n_f}$ the numerator. The characteristic polynomial, $v(q)$, of the output-feedback compensator, $F(q)$, is defined as the determinant of the denominator $V(q)$; i.e.,

$$v(q) \triangleq \det(V(q))\,.$$

The polynomial order, $n_v = \partial[v(q)]$, is the McMillan degree of $F(q)$.

$R(q) \in \Re(q)^{n_i \times n_i}$ is a proper (causal), pre-compensator, the inputs to which are the n_i reference signals. The design methodology requires $R(q)$ be described by a LMFD; i.e.,

$$R(q) = V^{-1}(q).M(q)$$

where $M(q) \in \Re[q]^{n_i \times n_i}$. The denominator $V(q)$ is identical to that of the output-feedback compensator, $F(q)$, thus the characteristic polynomial of $R(q)$ is $v(q)$.

$U \in \Re^{n_f \times n_o}$ is a matrix describing the interconnection between the n_o controlled outputs and the n_f feedback signals which are available to the feedback compensator.

Let $G_F(q) = U.N_R(q).D_R^{-1}(q) \in \Re(q)^{n_f \times n_i}$ be the modified plant, where $G_F(q)$ is of least order; the McMillan degree of $G_F(q)$ is that of $G(q)$. Further, let $S(A, B, C_F)$ denote the SSD of $G_F(q)$, where $C_F = U.C$.

For the closed-loop system shown in Fig. 4.1, the linear control law is thus given by

$$u(t) = R(q).r(t) - F(q).c_F(t), \tag{4.1}$$

where:

$u(t) \in \Re^{n_i}$ are the actuating (or error) input signals;

$r(t) \in \Re^{n_i}$ are the reference inputs signals;

$c_F(t) \in \Re^{n_f}$ are the feedback signals which are available to the feedback compensator; and

$c(t) \in \Re^{n_o}$ are the controlled outputs of the plant.

4.2.2 The Closed-Loop Transfer Function

Let us determine the input-output behaviour of the closed-loop system, assuming zero initial conditions. The closed-loop transfer function, $H(q) \in \Re(q)^{n_o \times n_i}$, relating the input-output behaviour of the reference inputs, $r(t)$, and the controlled outputs, $c(t)$, is given by

$$c(t) = H(q).r(t)\,.$$

Consider the linear control law

$$u(t) = R(q).r(t) - F(q).c_F(t)$$

of the closed-loop system. Substituting the LMFD of both the pre-compensator, $R(q)$, and the output-feedback compensator, $F(q)$, into the linear control law, we find

$$\begin{aligned} u(t) &= V^{-1}(q).M(q).r(t) - V^{-1}(q).K(q).c_F(t) \\ &= V^{-1}(q).M(q).r(t) - V^{-1}(q).K(q).U.c(t)\,. \end{aligned} \tag{4.2}$$

The RMFD of the plant, $G(q)$, is

$$c(t) = N_R(q).D_R^{-1}(q).u(t)\,. \tag{4.3}$$

Substitution of eqn. (4.2) into eqn. (4.3) yields:

$$I_{n_o}.c(t) = N_R(q).D_R^{-1}(q).[V^{-1}(q).M(q).r(t) - V^{-1}(q).K(q).U.c(t)], \qquad (4.4)$$

where I_{n_0} is the $n_o \times n_o$ identity matrix. Collecting like terms in eqn. (4.4), we have

$$(I_{n_o} + N_R(q).D_R^{-1}(q).V^{-1}(q).K(q).U).c(t) = N_R(q).D_R^{-1}(q).V^{-1}(q).M(q).r(t)$$

or, equivalently,

$$c(t) = (I_{n_o} + \overbrace{N_R(q).D_R^{-1}(q).V^{-1}(q).K(q).U}^{P.Q})^{-1}.\overbrace{N_R(q).D_R^{-1}(q)}^{P}.V^{-1}(q).M(q).r(t)\,. \qquad (4.5)$$

Consider the following matrix identity,

$$(I_{n_r} + P.Q)^{-1}.P = P.(I_{n_c} + Q.P)^{-1},$$

where P and Q are of dimension $n_r \times n_c$ and $n_c \times n_r$, respectively. Apply this identity to eqn. (4.5), with $P = N_R(q).D_R^{-1}(q)$ and $Q = V^{-1}(q).K(q).U$, the dimensions of which are $n_o \times n_i$ and $n_i \times n_o$, respectively. Hence eqn. (4.5) is equivalent to

$$c(t) = \overbrace{N_R(q).D_R^{-1}(q)}^{P}.(I_{n_i} + \overbrace{V^{-1}(q).K(q).U.N_R(q).D_R^{-1}(q)}^{Q.P})^{-1}.V^{-1}(q).M(q).r(t)\,. \qquad (4.6)$$

Using the matrix identity $Z^{-1}.Y^{-1}.X^{-1} = (X.Y.Z)^{-1}$, where $Z = D_R(q)$, $Y = (I_{n_i} + V^{-1}(q).K(q).U.N_R(q).D_R^{-1}(q))$ and $X = V(q)$, we find

$$\begin{aligned} c(t) &= N_R(q).\{V(q).[I_{n_i} + V^{-1}(q).K(q).U.N_R(q).D_R^{-1}(q)].D_R(q)\}^{-1}.M(q).r(t) \\ &= N_R(q).[V(q).D_R(q) + K(q).U.N_R(q)]^{-1}.M(q).r(t), \end{aligned}$$

or, equivalently,

$$c(t) = N_R(q).[V(q).D_R(q) + K(q).\bar{N}_R(q)]^{-1}.M(q).r(t),$$

where

$$\bar{N}_R(q) = U.N_R(q)\,.$$

Hence the closed-loop transfer function is of the form

$$H(q) = N_R(q).W^{-1}(q).M(q), \qquad (4.7)$$

where

$$W(q) = V(q).D_{\mathrm{R}}(q) + K(q).\bar{N}_{\mathrm{R}}(q) \tag{4.8}$$

is known as the *left compensator equation* (LCE).

Remark 4.1 The transfer function of the closed–loop system, $H(q)$, is composed of three polynomial matrices, these being, (i) the plant numerator, $N_{\mathrm{R}}(q)$, (ii) the closed–loop denominator, $W(q)$, and (iii) the pre–compensator numerator, $M(q)$. Let us consider the unassigned polynomial matrices $W(q)$ and $M(q)$:

1. The closed–loop denominator, $W(q)$, must be non–singular for the closed–loop transfer function to exist.

2. The right–hand side of the $n_{\mathrm{i}} \times n_{\mathrm{i}}$ LCE is composed of the polynomial matrices $\bar{N}_{\mathrm{R}}(q)$, $D_{\mathrm{R}}(q)$ of the MFDs of the modified plant, $G_F(q)$, and $V(q)$, $K(q)$ of the output–feedback compensator, $F(q)$, respectively. Since the modified plant, $G_F(q)$, and the output–feedback compensator, $F(q)$, are coprime representations, the closed–loop characteristic polynomial can be expressed as [16,91]

$$w(q) = \det(W(q)),$$

the polynomial order of which is $n_{\mathrm{w}} = \partial[w(q)]$. Thus the eigenvalues of $W(q)$ are the poles of the closed–loop system.

3. The LCE is independent of the numerator, $M(q)$, of the pre-compensator, $R(q)$. $M(q)$ thus provides an additional degree of freedom in the design process and is used to satisfy the steady–state requirements of the closed–loop system.

Result 4.1 The degrees of the denominators $D_{\mathrm{R}}(q)$, $V(q)$ and $W(q)$ of the modified plant, output–feedback compensator and closed–loop system, respectively, are related through the inequality

$$\partial[W(q)] \leq \partial[V(q)] + \partial[D_{\mathrm{R}}(q)]\,. \tag{4.9}$$

In general, the polynomial matrices $K(q)$ and $\bar{N}_{\mathrm{R}}(q)$ are non–square with

$$\partial[K(q)] \leq \partial[V(q)] \text{ and } \partial[\bar{N}_{\mathrm{R}}(q)] \leq \partial[D_{\mathrm{R}}(q)],$$

then

$$\partial[K(q).\bar{N}_{\mathrm{R}}(q)] \leq \partial[W(q)] \,.$$

The orders of $D_{\mathrm{R}}(q)$, $V(q)$ and $W(q)$ are related through the equality

$$n_{\mathrm{w}} = n_{\mathrm{v}} + n_{\mathrm{s}} \,. \tag{4.10}$$

The dynamic output–feedback compensator of order n_{v} thus introduces n_{v} additional poles in the resulting closed–loop system.

4.3 The Output–Feedback Compensator

4.3.1 The Modified Left Compensator Equation

Freedom in pole placement is the basis for designing the dynamics of the closed–loop system. The problem of pole assignment in MIMO systems using dynamic output–feedback compensation can be stated as follows. Given the modified plant, $G_F(q) = \bar{N}_{\mathrm{R}}.D_{\mathrm{R}}^{-1}(q)$, and a set of desired closed–loop poles specified through $W(q)$, determine the dynamic output–feedback compensator $F(q) = V^{-1}(q).K(q)$ such that eqn. (4.8) is satisfied.

It is well known (Brasch and Pearson [14], Pearson and Ding [95], Chen and Hsu [19], Chen [18]) that for a plant, $G_F(q) = \bar{N}_{\mathrm{R}}(q).D_{\mathrm{R}}^{-1}(q)$, it is always possible to design dynamic output–feedback compensators such that the poles of the closed–loop system can be positioned arbitrarily. Recall that n_{s}, n_{f} and n_{i} are the plant order, and the numbers of feedback signals and inputs, respectively. It has been shown that, by use of zero order (static) output feedback, as many as $\min(n_{\mathrm{s}}, n_{\mathrm{f}} + n_{\mathrm{i}} - 1)$ system poles can be assigned arbitrarily, for almost all plants, provided a slight modification of the pole locations to be assigned are tolerable (Davison and Wang [23], Kimura [61,62]). However, arbitrary assignment of all closed–loop poles using static output feedback is not feasible for $n_{\mathrm{s}} > n_{\mathrm{f}} + n_{\mathrm{i}} - 1$. In this case, a dynamic output–feedback compensator is required for the arbitrary assignment of more than $(n_{\mathrm{f}} + n_{\mathrm{i}} - 1)$ poles. It has been shown that a dynamic output–feedback compensator of order n_{v} can assign $\min(n_{\mathrm{s}}+n_{\mathrm{v}}, n_{\mathrm{f}}+n_{\mathrm{i}}-1+2n_{\mathrm{v}})$

closed–loop poles, for almost all plants, provided a slight modification of the pole locations are tolerable [23].

Let us formally introduce the above results.

Theorem 4.1 [23]: Given an observable and controllable SSD $S(A, B, C_F)$ of a plant with $A \in \Re^{n_s \times n_s}$, rank$(B) = n_i$, rank$(C_F) = n_f$, then for almost all (B, C_F) pairs, there exists a zero order (static) output–feedback compensator so that the closed–loop system has

$$\min(n_s, n_i + n_f - 1) \tag{4.11}$$

poles assigned arbitrarily close to $\min(n_s, n_i + n_f - 1)$ specified poles.

Remark 4.2 It should be noted that the methodology which will be proposed for pole placement assigns the specified closed–loop poles exactly rather than 'arbitrarily close', and thus the result of Theorem 4.1 may not be realised.

Result 4.2 If the plant $S(A, B, C_F)$ does not satisfy the condition

$$n_s \leq n_i + n_f - 1,$$

then using a dynamic output–feedback compensator of appropriate order, it is possible to assign more than $(n_f + n_i - 1)$ closed–loop poles. As was shown by Brasch and Pearson [14], the problem of pole–assignment for the modified plant $S(A, B, C_F)$ using a dynamic output–feedback compensator can be reduced to the problem of pole–assignment for the SSD $S(\hat{A}, \hat{B}, \hat{C}_F)$ using static output feedback, where a set of n_v intergrators is adjoined to the plant to give the following augmented system.

$$\hat{A} = \begin{bmatrix} A & 0_{n_s \times n_v} \\ 0_{n_v \times n_s} & 0_{n_v \times n_v} \end{bmatrix}, \hat{B} = \begin{bmatrix} B & 0_{n_s \times n_v} \\ 0_{n_v \times n_i} & I_{n_v \times n_v} \end{bmatrix}, \hat{C}_F = \begin{bmatrix} C_F & 0_{n_f \times n_v} \\ 0_{n_v \times n_s} & I_{n_v \times n_v} \end{bmatrix}.$$

Replacing the numbers n_s, n_i, n_f by $n_s + n_v$, $n_i + n_v$, $n_f + n_v$ in eqn. (4.11), respectively, then for a dynamic output–feedback compensator of order n_v we can assign

$$\min(n_s + n_v, n_f + n_i - 1 + 2n_v) \tag{4.12}$$

poles arbitrarily close to the same number of specified closed–loop poles, for almost all plants.

From the above result, a feedback compensator of order $n_v = n_s - n_f - n_i + 1$ can assign, for almost all plants, the $n_s + n_v$ closed–loop poles arbitrarily close to their desired values.

$\nabla\nabla\nabla$

The above results provide the designer with a lower–bound estimate only of the order of the output–feedback compensator required to arbitrarily assign a specified number of closed–loop poles. Since the methodology proposed for pole–placement assigns the closed–loop poles exactly the compensator order may, in fact, be greater than that given by the above results.

Definition 4.1 The poles arbitrarily assigned by a designer are defined as *designer–assigned* poles. The poles assigned in the process of solving the LCE are defined as *solution–assigned* poles.

Proposition 4.1 The form of the LCE, eqn. (4.8), does not provide the designer the flexibility of choosing the number of designer–assigned and solution–assigned closed–loop poles. Moreover, in some applications the designer might constrain the order of the output–feedback compensator to a specific value, or might not wish to assign all closed–loop poles. In such cases, complete freedom in pole assignment is no longer applicable. To accommodate these situations the LCE is modified as follows:

$$W_s(q).W_d(q) = V(q).D_R(q) + K(q).\bar{N}_R(q), \tag{4.13}$$

where the closed–loop denominator, $W(q)$, is described by the product

$$W_s(q).W_d(q)$$

of two $n_i \times n_i$ non–singular polynomial matrices. The eigenvalues of the polynomial matrices $W_s(q)$ and $W_d(q)$ are associated with the solution–assigned and designer–assigned pole positions of the closed–loop system, respectively.

Let us denote the modified left compensator equation (4.13) by MLCE.

Result 4.3 The characteristic polynomial, $w(q)$, of the closed-loop system can be expressed as the product of two polynomials

$$w_s(q).w_d(q),$$

where the zeros of $w_s(q)$ and $w_d(q)$ are the solution-assigned and designer-assigned closed-loop poles, respectively.

Result 4.4 The relationship between the order of the modified plant, the output-feedback compensator and the closed-loop system of eqn. (4.10) is modified to take into account the orders of solution-assigned and designer-assigned closed-loop poles, i.e.,

$$n_w = n_{w_d} + n_{w_s} \tag{4.14}$$

and

$$n_{w_s} + n_{w_d} = n_v + n_s \,. \tag{4.15}$$

▽▽▽

It should be noted that the solution-assigned closed-loop poles can move to undesirable, even unstable, locations. Though there is no requirement that the output-feedback compensator, $F(q)$, be stable, it will be assumed in the examples in Chapter 5 that this is the case. Solution-assigned closed-loop poles must, of course, be stable.

4.3.2 Solving the Modified Left Compensator Equation

In this section the problem of solving the MLCE (4.13) will be considered. It is possible to cast this problem into a standard form involving the solution of the polynomial matrix equation [40,115]:

$$C(q).P(q) = W(q), \tag{4.16}$$

where

$$C(q) = \begin{bmatrix} K(q) & V(q) \end{bmatrix} \text{ and } P(q) = \begin{bmatrix} \bar{N}_R \\ D_R(q) \end{bmatrix}. \tag{4.17}$$

By imposing an additional constraint that $F(q)$ should have McMillan degree, we ensure that the output-feedback compensator has minimal dynamic order. In the literature the problem of solving equations of the form of eqn. (4.16), with specified constraints, has been called the Minimal Design Problem (MDP) [42,58,64,80,105,115]. This problem arises in this and several other areas of linear system theory, e.g. in exact model matching and inverse systems [40,42,57,58,64,80,105,115].

The MDP can be cast into a problem of determining a *polynomial basis* for a vector space over the ring of polynomials.

Let us now rearrange eqn. (4.16) to form the following MLCE,

$$\begin{bmatrix} W_s(q) & K(q) & V(q) \end{bmatrix} . \begin{bmatrix} -W_d(q) \\ \bar{N}_R(q) \\ D_R(q) \end{bmatrix} = 0 \,. \tag{4.18}$$

This may be expressed as

$$Y(q).X(q) = 0, \tag{4.19}$$

where

$$Y(q) = \begin{bmatrix} W_s(q) & C(q) \end{bmatrix} \tag{4.20}$$

is a $n_i \times (\bar{n} = n_f + 2n_i)$ polynomial matrix and

$$X(q) = \begin{bmatrix} -W_d(q) \\ P(q) \end{bmatrix} \tag{4.21}$$

is a $\bar{n} \times n_i$ polynomial matrix having full column rank n_i. Further, it is assumed that the polynomial matrices $W_d(q)$ and $P(q)$ are known or specified. The problem of solving for $Y(q)$ in eqn. (4.19) is equivalent to that of finding a basis in the left null space of the polynomial matrix $X(q)$, that is, a solution for $Y(q)$ is contained within the null space of $X(q)$. Furthermore, since we are interested in solutions of least possible order, we must search for a minimal basis in the null space. The requirement that $F(q)$ be proper (causal) imposes a further constraint on the basis of the left null space of $X(q)$.

We shall now describe a method for solving eqn. (4.18) for the polynomial matrices $W_s(q)$, $K(q)$ and $V(q)$. At first glance, it might seem that any method for solving eqn. (4.18) would involve using operations on polynomial matrices and would therefore be quite involved and computationally rather unattractive. However, as we shall see, this is not the case. In order to avoid the complexity of dealing with polynomial matrices, we reduce the problem to one in which we use the method of resultants (see § 2.4.4) to transform the MLCE into a set of linear equations. We can then operate with matrices of real numbers rather than with polynomial matrices, thereby making the method computationally attractive.

It is clear from our definition of the MDP that the solution of eqn. (4.18) must satisfy the following conditions:

1. $V(q)$ and $K(q)$ must be *left–coprime* polynomial matrices;
2. $F(q) = V^{-1}(q).K(q)$ must be a *proper (causal)* rational function matrix;
3. the polynomial order, n_{v}, of $\det(V(q))$ must be as small as possible; and
4. the polynomial order, $n_{\mathrm{w_s}}$, of $\det(W_{\mathrm{s}}(q))$ must equal $n_{\mathrm{w_s}} = n_{\mathrm{s}} + n_{\mathrm{v}} - n_{\mathrm{w_d}}$.

These four conditions impose constraints on the solutions of eqn. (4.18). We must examine this equation and its solutions more closely to determine explicitly if and when the above conditions are satisfied.

It is clear from eqn. (4.19) that the polynomial matrix $Y(q)$ is contained in the left null space of $X(q)$. Therefore, by obtaining a minimal basis (see Definition 2.5) for this left null space, or left minimal basis (LMB) for short, we can generate all possible solutions of eqn. (4.19). Some of these solutions may satisfy the four constraints listed above, in which event we may have a solution for the MDP.

Let us introduce the following result which is required later.

Result 4.5 [4]: The dimension of the left null space (or the *left nullity*) of a polynomial matrix $L(q) \in \Re^{n_{\mathrm{r}} \times n_{\mathrm{c}}}[q]$, $n_{\mathrm{r}} \geq n_{\mathrm{c}}$, is

$$(\text{left nullity of } L(q))+(\text{rank of } L(q)) = n_{\mathrm{r}}\,.$$

Further, if $L(q)$ is of full column rank then (rank of $L(q)$)=n_{c}. Thus, it follows that the dimension of the left null space of a full column–rank polynomial matrix $L(q)$ is equal to the difference between its row and column dimensions; i.e., $n_{\mathrm{r}} - n_{\mathrm{c}}$. Note also, that the rank of the left null space is equal to its left nullity.

$\nabla\nabla\nabla$

Let us denote the LMB of $X(q)$ by

$$\tilde{Y}(q) = \left[\begin{array}{ccc} \tilde{W}_s(q) & \tilde{K}(q) & \tilde{V}(q) \end{array}\right] \tag{4.22}$$

such that

$$\tilde{Y}(q).X(q) = 0\,. \tag{4.23}$$

With reference to Result 4.5, the dimension of the left null space, $\tilde{Y}(q)$, of $X(q)$ is $\bar{n} - n_i = n$. Thus, $\tilde{Y}(q)$ is a $(n = n_f + n_i) \times (\bar{n} = n_f + 2n_i)$ polynomial matrix having full row rank n; furthermore, $\tilde{W}_s(q)$, $\tilde{K}(q)$, $\tilde{V}(q)$ are $n \times n_i$, $n \times n_f$ and $n \times n_i$ polynomial matrices, respectively.

By Definition 2.5 (of a minimal basis) and Definition 2.7 (of a row–proper matrix), it is easy to see that $\tilde{Y}(q)$ satisfies the following conditions:

1. The greatest common divisor of $n \times n$ minors of $\tilde{Y}(q)$ is 1; i.e., it has only unimodular (left) divisors.

2. It is row proper; i.e., its highest row–degree coefficient matrix $\Gamma_r[\tilde{Y}(q)]$ has full rank.

As a result of these two conditions, a minimal basis satisfies the first constraint imposed on the MDP, i.e., $V(q)$ and $K(q)$ must be *left–coprime* polynomial matrices.

We now turn our attention to the second constraint imposed on the MDP, i.e., $F(q)$ must be a *proper (causal)* rational function matrix. The following result gives a necessary and sufficient condition for the existence of a proper (causal) solution to the output–feedback compensator. However, let us first introduce the following proposition.

Proposition 4.2 Let

$$\widehat{C}(q) = \left[\begin{array}{cc} \widehat{K}(q) & \widehat{V}(q) \end{array}\right]$$

be a full row rank $\hat{n} \times n$ polynomial matrix, the rows of which are those of the submatrix $[\tilde{K}(q)\ \tilde{V}(q)]$ of $\tilde{Y}(q)$ after the linearly dependent rows have been extracted, where $\hat{n} = \mathrm{rank}([\tilde{K}(q)\ \tilde{V}(q)])$ with $\widehat{K}(q)$ and $\widehat{V}(q)$ being of dimension $\hat{n} \times n_f$ and $\hat{n} \times n_i$, respectively.

Let

$$\widehat{Y}(q) = \left[\begin{array}{ccc} \widehat{W}_s(q) & \widehat{K}(q) & \widehat{V}(q) \end{array}\right], \tag{4.24}$$

where $\widehat{W}_s(q)$ is of dimension $\hat{n} \times n_i$, the rows of which are those of $\tilde{W}_s(q)$ in $\tilde{Y}(q)$ which occupy the same rows as the $\hat{n}$ linearly independent rows of the submatrix $[\tilde{K}(q)\ \tilde{V}(q)]$ of $\tilde{Y}(q)$.

If $\widehat{C}(q)$ is *not* row-proper, then premultiply $\widehat{C}(q)$ and $\widehat{Y}(q)$ by a unimodular matrix (see § 2.4.9) which reduces $\widehat{C}(q)$ to its row-proper form.

Theorem 4.2 [92]: Let

$$\widehat{C}(q) = \left[\ \widehat{K}(q)\ \ \widehat{V}(q)\ \right]$$

be a full row rank, row-proper matrix and let its highest row-degree coefficient matrix $\Gamma_r[\widehat{C}(q)]$ be partitioned as

$$\Gamma_r[\widehat{C}(q)] = \left[\ \widehat{K}_\Gamma\ \ \widehat{V}_\Gamma\ \right] \tag{4.25}$$

where $\widehat{K}_\Gamma$ and $\widehat{V}_\Gamma$ are $\hat{n} \times n_f$ and $\hat{n} \times n_i$, respectively. Then a proper (causal) solution to the MDP exists if and only if

$$\text{rank}(\widehat{V}_\Gamma) = n_i\,. \tag{4.26}$$

$\nabla\nabla\nabla$

Let us consider the following example to illustrate Theorem 4.2.

Example 4.1 Consider the polynomial matrix

$$\widehat{C}(q) = \left[\ \widehat{K}(q)\ \ \widehat{V}(q)\ \right] = \left[\begin{array}{cc|cc} 2q^2 & 4q & q & 3 \\ 1 & 3 & 1 & 9q \end{array}\right].$$

The highest row-degree coefficient matrix $\Gamma_r[\widehat{C}(q)]$ is

$$\Gamma_r[\widehat{C}(q)] = \left[\ \widehat{K}_\Gamma\ \ \widehat{V}_\Gamma\ \right] = \left[\begin{array}{cc|cc} 2 & 0 & 0 & 0 \\ 0 & 0 & 0 & 9 \end{array}\right]$$

which has full rank, and thus $\widehat{C}(q)$ is row proper. Note that $\widehat{V}_\Gamma$ does not have full rank. However, the polynomial matrix $\widehat{V}(q)$ has row indices $\mu_1 = 1$ and $\mu_2 = 1$ and

$$\Gamma_r[\widehat{V}(q)] = \begin{bmatrix} 1 & 0 \\ 0 & 9 \end{bmatrix}.$$

Therefore $\text{rank}(\Gamma_r[\widehat{V}(q)]) = 2$ and $\widehat{V}(q)$ is row-proper.

It should be noted that $\widehat{V}_\Gamma$ and $\Gamma_r[\widehat{V}(q)]$ may not necessarily be the same. In fact, they may not even have the same rank (as in this example). It can be easily shown that, for a row–proper matrix $\widehat{C}(q)$, if $\widehat{V}_\Gamma$ has full rank then so has $\Gamma_r[\widehat{V}(q)]$ but the converse is not always true (as has just been noted). In this example a proper (causal) solution for the output–feedback compensator does not exist, which is in agreement with Theorem 4.2.

$\nabla\nabla\nabla$

Let us now investigate how the third of the constraints imposed on the MDP can be satisfied, i.e., the polynomial order, n_v, of $\det(V(q))$ must be as small as possible. For this a solution of eqn. (4.18) is required which gives rise to a coprime rational output–feedback compensator, $F(q) = V^{-1}(q).K(q)$, having least order. In order to find such a solution, it is assumed, without loss of generality, that the $\hat{n}$ rows of $\widehat{Y}(q)$ are arranged such that the row indices, $\hat{\mu}_j$ $j \in \underline{\hat{n}}$, of $\widehat{C}(q)$ are in ascending order (from top to bottom); i.e.,

$$\hat{\mu}_1 \leq \hat{\mu}_2 \leq \ldots \leq \hat{\mu}_{\hat{n}} \,. \tag{4.27}$$

Then, proceeding from top to bottom, we choose those rows of $\widehat{Y}(q)$ which correspond to the first n_i linearly independent rows of $\widehat{V}_\Gamma$. This yields the matrix $Y(q)$ which can be partitioned as follows,

$$Y(q) = \left[\begin{array}{ccc} W_s(q) & K(q) & V(q) \end{array} \right] . \tag{4.28}$$

We then have the following left–coprime (i.e., having McMillan degree), proper (causal), least–order solution to the MDP for the output–feedback compensator

$$F(q) = V^{-1}(q).K(q) \,. \tag{4.29}$$

The fourth constraint of the MDP is immediately satisfied when constraints (i)-(iii) have been satisfied. This follows directly from Item 2 of Remark 4.1 and eqns. (4.14) and (4.15) of Result 4.4.

Proposition 4.3 The output–feedback compensator of eqn. (4.29) is a feasible solution to the MDP.

Proof 4.1 It is clear that Proposition 4.3 is true from the following:

1. The polynomial matrix $Y(q)$, having only unimodular (left) divisors, lies in the left null space of $X(q)$.

2. The condition that rank($\hat{V}_\Gamma$) $= n_\text{i}$ implies, according to Theorem 4.2, that the solution $Y(q)$ gives rise to a proper (causal) rational output–feedback compensator $V^{-1}(q).K(q)$.

3. The rows of $Y(q)$ are chosen from those of $\hat{Y}(q)$ so as to yield a left–coprime, proper (causal), least–order solution for $F(q)$. Since rank($\hat{V}_\Gamma$) $= n_\text{i}$, and the row indices of $V(q)$ are equal to the row indices of $\hat{C}(q)$, it follows that the polynomial order, n_v, of det($V(q)$) is equal to the sum of these row indices.

4. The condition $n_{\text{w}_\text{s}} = n_\text{v} + n_\text{d} - n_{\text{w}_\text{d}}$ immediately follows when the above three conditions are satisfied.

$\nabla\nabla\nabla$

So far we have examined the various conditions under which the MDP has a solution. We have assumed in the discussion that a LMB $\tilde{Y}(q)$ of $X(q)$ has been derived by some means. The determination of such a basis is not trivial. Several methods exist for obtaining a minimal basis or for reducing a nonminimal basis to a minimal one (see for example [40,42,57,58,64,80,105,115]). A method [42,57,58,64] is now described which has some advantages from the point of view of computation. We begin by introducing the concept of a "Popov echelon form" for a polynomial matrix.

Notation: For notational convenience assume that the jth column of $\tilde{Y}(q)$ or $\tilde{Y}$ ($\tilde{Y}$ is not yet defined) is taken with reference to its right–most column. Likewise, assume that the ith row of $X(q)$, $\mathcal{X}(q)$, or X_R ($\mathcal{X}(q)$ and X_R are not yet defined) is taken with reference to its bottom–most row.

Definition 4.2 [57,58]: A $n \times \bar{n}$ polynomial matrix $\tilde{Y}(q)$ is said to be in *Popov echelon form* if it satisfies the following conditions:

1. $\tilde{Y}(q)$ is row–proper with its *row indices* in ascending order; i.e.,

$$\mu_1 \leq \mu_2 \leq \ldots \leq \mu_n \, .$$

2. For row i, there exists an index ρ_i, called a *pivot index*, such that

(a) $\tilde{Y}_{i\rho_i}(q)$ has degree μ_i.

(b) $\tilde{Y}_{i\rho_i}(q)$ is monic.

(c) $\tilde{Y}_{i\rho_i}(q)$ is the last entry (i.e., left–most) in the ith row having degree μ_i; i.e., $\partial[\tilde{Y}_{ij}(q)] < \mu_i$ if $j > \rho_i$ (note that the jth column of $\tilde{Y}(q)$ is taken with reference to its right–most column).

(d) If $\mu_i = \mu_j$ and $i < j$, then $\rho_i < \rho_j$; i.e., the pivot indices are in ascending order.

(e) The pivot indices ρ_i, $i \in \underline{n}$, are distinct ($\underline{k}$ is the set of integers $(1, 2, \ldots, k)$).

(f) $\partial[\tilde{Y}_{j\rho_i}(q)] < \mu_i$ if $j \neq i$, $j \in \underline{n}$.

$\nabla\nabla\nabla$

The Popov echelon form as defined by the above conditions is unique (i.e. canonical).

On examination of a polynomial matrix in Popov echelon form, it is not immediately obvious that there is any specific "echelon" structure associated with the matrix. This structure, however, becomes apparent when we consider the row block–matrix description of the given polynomial matrix. To illustrate this, consider the following example.

Example 4.2 It can be easily verified that the polynomial matrix given by

$$\tilde{Y}(q) = \begin{bmatrix} 5 & \underline{q+7} & 4q+6 & 2q+4 & 3q+1 \\ 7q+2 & 8 & 3q+6 & q+4 & \underline{q^2+6q+2} \\ \underline{q^2+2q+6} & 4 & 3q^2+8q+3 & 4q^2+2q+3 & 2q+4 \\ 6q+4 & 5 & \underline{q^3+q^2+7} & 4q^3+7q^2+q+7 & 2q+3 \end{bmatrix} \tag{4.30}$$

is in Popov echelon form.

Note that $n = 4$, $\bar{n} = 5$, $\mu_1 = 1$, $\mu_2 = 2$, $\mu_3 = 2$ and $\mu_4 = 3$, and the pivot indices are $\rho_1 = 4$, $\rho_2 = 1$, $\rho_3 = 5$ and $\rho_4 = 3$ (the corresponding elements of $\tilde{Y}(q)$ are underlined). The row block–matrix description $\tilde{Y}$ of $\tilde{Y}(q)$ is given by

$$\tilde{Y} = \left[\begin{array}{ccccc|ccccc|ccccc|ccccc} 0&0&0&0&0 & 0&0&0&0&0 & 0&\underline{1}&4&2&3 & 5&7&6&4&1 \\ 0&0&0&0&0 & 0&0&0&0&\underline{1} & 7&0&3&1&6 & 2&8&6&4&2 \\ 0&0&0&0&0 & \underline{1}&0&3&4&0 & 2&0&8&2&2 & 6&4&3&3&4 \\ 0&0&\underline{1}&4&0 & 0&0&1&7&0 & 6&0&0&1&2 & 4&5&7&7&3 \end{array}\right]. \tag{4.31}$$

It should be noted that $\tilde{Y}$ is in the standard reduced row echelon form for real matrices; the pivot entries have been underlined in the above matrix.

However, the Popov echelon form imposes an additional constraint on the pivot entries in $\tilde{Y}$. Let the pivot entry in the ith row of $\tilde{Y}$ be in column η_i. Then the set of integers $\{\eta_i,\ i \in \underline{n}\}$ satisfy the following condition

$$\eta_i \neq \eta_j \bmod \bar{n} \text{ if } i \neq j \ . \tag{4.32}$$

$\nabla\nabla\nabla$

Let us now consider a methodology described by Kung and Kailath [64] for determining a LMB $\tilde{Y}(q)$ of $X(q)$ in the Popov echelon form, satisfying

$$\tilde{Y}(q).X(q) = 0\,. \tag{4.33}$$

Denote the degrees of $\tilde{Y}(q)$ and $X(q)$ by

$$p = \partial[\tilde{Y}(q)] = \max(\partial[\tilde{W}_s(q)], \partial[\tilde{V}(q)]) \tag{4.34}$$

and

$$r = \partial[X(q)] = \max(\partial[W_d(q)], \partial[D_R(q)]), \tag{4.35}$$

respectively. The degree p of $\tilde{Y}(q)$ is usually not known *a priori*, but we shall assume a lower bound given by

$$p_{\min} = \max(\partial[V(q)]_{\min}, \partial[W_s(q)]_{\min}), \tag{4.36}$$

where $\partial[V(q)]_{\min}$ and $\partial[W_s(q)]_{\min}$ denote the smallest possible degrees that can be assigned to the polynomial matrices $V(q)$ and $W_s(q)$ of dimension n_i and polynomial order n_v and n_{w_s}, respectively. The degrees $\partial[V(q)]_{\min}$ and $\partial[W_s(q)]_{\min}$ are determined through the computation of the minimal degree which satisfies the inequality of eqn. (2.2) of Result 2.2. Note that the resulting degree, p, of $\tilde{Y}(q)$ may not necessarily equal $p_{\min}$, that is, p may increase in the course of the computation.

Comment: Note, in the row block–matrix description of $\tilde{Y}(q)$, the block–matrix $\tilde{Y}_j$ corresponding to the indeterminate of degree $p - j$, $j = 0, \ldots, p$, is

$$\tilde{Y}_j = \begin{bmatrix} \tilde{W}_{s_j} & \tilde{K}_j & \tilde{V}_j \end{bmatrix},$$

where, for $\tilde{W}_s(q)$, $\tilde{K}(q)$ and $\tilde{V}(q)$, respectively, the matrices $\tilde{W}_{s_j}$, $\tilde{K}_j$ and $\tilde{V}_j$ denote the block–matrices corresponding to the indeterminate of degree $p{-}j$. Similarly, in the column block–matrix description of $X(q)$, the block–matrix X_j corresponding to the indeterminate of degree $r-j$, $j=0,\ldots,r$, is

$$X_j = \begin{bmatrix} -W_{\mathrm{d}_j} \\ \bar{N}_{\mathrm{R}_j} \\ D_{\mathrm{R}_j} \end{bmatrix}.$$

$\nabla\nabla\nabla$

Let us now apply the method of resultants to eqn. (4.33) (see § 2.4.5), the numerical representation of which is written as

$$\tilde{Y}.(< X >_p = X_{\mathrm{R}}) = 0\,, \tag{4.37}$$

that is:

$$\begin{bmatrix} \tilde{Y}_0 & \tilde{Y}_1 & \ldots & \tilde{Y}_p \end{bmatrix} . \begin{bmatrix} X_0 & X_1 & \ldots & \ldots & X_r & 0 & \ldots & \ldots & 0 \\ 0 & X_0 & X_1 & \ldots & \ldots & X_r & \ddots & & \vdots \\ \vdots & \ddots & \ddots & \ddots & & & \ddots & \ddots & \vdots \\ \vdots & & \ddots & \ddots & \ddots & & & \ddots & 0 \\ 0 & \ldots & \ldots & 0 & X_0 & X_1 & \ldots & \ldots & X_r \end{bmatrix} = 0\,. \tag{4.38}$$

Remark 4.3 This result could also have been derived using the conventional approach based on polynomial matrices.

Using the polynomial matrix description for $\tilde{Y}(q)$,

$$\tilde{Y}(q) = \tilde{Y}_0 q^p + \tilde{Y}_1 q^{p-1} + \ldots + \tilde{Y}_p, \tag{4.39}$$

(as described by eqn. (2.1) of Definition 2.2), we can write eqn. (4.33) in the form

$$\begin{bmatrix} \tilde{Y}_0 & \tilde{Y}_1 & \ldots & \tilde{Y}_p \end{bmatrix} . \begin{bmatrix} X(q)q^p \\ X(q)q^{p-1} \\ \vdots \\ X(q) \end{bmatrix} = 0 \tag{4.40}$$

or, equivalently,

$$\tilde{Y}.\mathcal{X}(q) = 0, \tag{4.41}$$

where

$$\tilde{Y} = \begin{bmatrix} \tilde{Y}_0 & \tilde{Y}_1 & \ldots & \tilde{Y}_p \end{bmatrix} \tag{4.42}$$

denotes the row block–matrix description of $\tilde{Y}(q)$ and

$$\mathcal{X}(q) = \begin{bmatrix} X(q)q^p \\ X(q)q^{p-1} \\ \vdots \\ X(q) \end{bmatrix} . \tag{4.43}$$

▽▽▽

Let us suppose that $\tilde{Y}(q)$ is in the Popov echelon form, that is, $\tilde{Y}$ has a reduced row echelon structure, and let us examine the constraints that such a structure imposes on X_R (– or $\mathcal{X}(q)$) through eqn. (4.38). This enables us to describe a methodology which can be used to locate the pivots of $\tilde{Y}$ (– or of $\tilde{Y}(q)$) and to construct the row block–matrix $\tilde{Y}$ in the Popov echelon form. Note from the earlier examination of the pivots in $\tilde{Y}$, they are defined to be located in columns η_i, $i \in \underline{n}$. Then, from eqn. (4.38) and the structure of $\tilde{Y}$, it is clear that each of the rows η_i, $i \in \underline{n}$, of X_R is a linear combination of the preceding rows (recall that the η_ith row of X_R is taken with reference to its bottom–most row), i.e.

$$X_{\mathrm{R}\,\eta_i} = aX_{\mathrm{R}\,(\eta_i-1)} + bX_{\mathrm{R}\,(\eta_i-2)} + \ldots + gX_{\mathrm{R}\,1} \,, \tag{4.44}$$

$$\begin{bmatrix} 1 & -a & -b & \ldots & -g \end{bmatrix} . \begin{bmatrix} X_{\mathrm{R}\,\eta_i} \\ \vdots \\ X_{\mathrm{R}\,1} \end{bmatrix} = 0 \,,$$

or,

$$\begin{bmatrix} 0_{t-\eta_i} & 1 & -a & -b & \ldots & -g \end{bmatrix} . \begin{bmatrix} X_{\mathrm{R}\,t} \\ \vdots \\ X_{\mathrm{R}\,1} \end{bmatrix} = 0 \,,$$

where t is the number of rows in the resultant X_{R} and 0_k is a vector of k null elements. The vector

$$\begin{bmatrix} 0_{t-\eta_i} & 1 & -a & -b & \ldots & -g \end{bmatrix} ,$$

is then the η_ith row of $\tilde{Y}$. Based on the notation of [57,58,64], these linearly dependent rows of X_R are called the *primary dependent rows* in order to distinguish them from the other dependent rows. Further, these rows η_i, $i \in \underline{n}$, are distinct in a strong sense, namely,

$$\eta_i \neq \eta_j \bmod \bar{n} \text{ if } i \neq j \; (\bar{n} = n_{\mathrm{f}} + 2n_{\mathrm{i}}).$$

We shall now examine the other dependent rows which will be known as *secondary (or nonprimary) dependent rows* [57,58]. Note that the polynomial matrix X_R has a special 'shift–invariant' block structure. These secondary dependent rows arise from the structure of X_R, namely, that $\mathcal{X}(q)$ is composed of $X(q)$ shifted by powers of q. It is clear that the sth row of X_R where

$$s = \eta_i \bmod \bar{n} \text{ for } s > \eta_i,\ i \in \underline{n} \tag{4.45}$$

is, according to the shift–invariance property of X_R, also a dependent row (because it is a linear combination of the preceding rows of X_R). Such rows satisfying eqn. (4.45) are called secondary dependent rows. We should note that only the primary dependent rows of X_R are useful in determining the solution block matrix $\tilde{Y}$ (and hence the polynomial matrix $\tilde{Y}(q)$). These primary dependent rows of X_R, can be obtained by the following *selection algorithm.*

Algorithm 4.1 [58,64]:

step 1. Set $i = 1$, $p = p_{\min}$ (evaluated from eqn. (4.36)).

step 2. Find the ith row, say η_i, of X_R that is linearly dependent on the preceding rows (the η_ith row of X_R being taken from its bottom–most row).

step 3. If a linearly dependent row is not found in step 2, set $p = p + 1$ and go to step 2; otherwise go to step 4.

step 4. Express the η_ith row of X_R as a linear combination of the preceding rows, that is, find the linear dependence coefficients $\tilde{Y}_{ij}$ such that

$$\sum_{j=1}^{\eta_i} \tilde{Y}_{ij}.X_{Rj} = 0 \tag{4.46}$$

where X_{Rj}, $j \in \underline{\eta_i}$, denotes the jth row of X_R and $\tilde{Y}_{ij}$, $i \in \underline{n}$, $j \in \underline{\eta_i}$, denotes the ith row, jth column element of $\tilde{Y}$.

step 5. Eliminate from the calculation row η_i and all rows $s(> \eta_i)$ of X_R, such that $s = \eta_i \bmod \bar{n}$.

step 6. If $i = n = (n_{\mathrm{i}} + n_{\mathrm{f}})$, stop; otherwise set $i = i + 1$ and go to step 2.

Remark 4.4 The aim is to find the primary dependent rows of X_R (– or $\mathcal{X}(q)$) in an efficient manner. This is done in step 4 using Gaussian elimination, say, operating on the resultant.

Remark 4.5 Algorithm 4.1 terminates when n primary dependent rows have been found. This implies that the value of $\partial[\tilde{Y}(q)]$ should be large enough to enable n primary dependent rows of $X(q)$ to be found. If the initial choice of $p_{\min}$ does not yield n primary dependent rows in $X(q)$, then the value of $p = \partial[\tilde{Y}(q)]$ is increased until n primary dependent rows are found. This procedure ensures that the value of p is as small as possible. Note that increasing the degree p is equivalent to augmenting X_R (– or $\mathcal{X}(q)$) by an appropriate number of block–rows. The selection rule in the algorithm clearly ensures that the resulting primary dependent rows have distinct pivot indices (modulo $\bar{n}$), and that all secondary dependent rows are excluded. The elements of $\tilde{Y}_{ij}$, $j = 1, \ldots, \eta_i$, in eqn. (4.46) represent the dependence relationships between the primary dependent rows and the rows preceding them. It is quite easy to see that the methodology ensures that $\tilde{Y}$ has the required reduced row echelon form, thus ensuring that $\tilde{Y}(q)$ is in Popov echelon form. It turns out that this construction yields an irreducible polynomial matrix, i.e., it has only left unimodular divisors [113].

To illustrate the above results the reader is referred to Example 4.3 in the final section of this chapter.

Algorithm 4.2 The following is the methodology proposed for the design of an output–feedback compensator, $F(q)$, using pole assignment.

step 1. For the plant, determine:

- the number of controlled outputs, n_{o};
- the number of reference inputs, n_{i};
- the number of plant poles, n_{s}, and their locations in the complex s– or z–plane.

step 2. If the plant is not represented in state–space form, then transform it to its equivalent observable and controllable state–space description.

step 3. If the solution for a discrete–time feedback compensator is required and if the plant SSD $S(A,B,C)$ is described in the continuous–time domain, then the plant is transformed to its discrete–time form. By application of the expressions [17]

$$A_D = e^{A.T} \text{ and } B_D = A^{-1}.(A_D - I_{n_s}).B, \tag{4.47}$$

the discrete–time, state–space equivalent (A_D, B_D, C) with zero–order hold can be realised, given a sampling period T. Commercial software routines are available for this purpose.

step 4. Form the RMFD of the plant by applying Algorithm 3.1.

step 5. Determine which of the n_o plant outputs are available for feedback. The modified plant $G_F(q) = \bar{N}_R(q).D_R^{-1}(q)$ is then constructed. Note that $G_F(q)$ must be right–coprime.

step 6. Select the number of closed–loop poles and their locations in the complex s– or z–plane that are to be arbitrarily assigned by the designer.

step 7. Using the relationship

$$n_{w_d} = \min(n_s + n_v, n_f + n_i - 1 + 2n_v),$$

select the order n_v of the output–feedback compensator to yield the desired number of closed–loop poles that are to be designer–assignable.

step 8. Construct the standard form, column structure matrix $P_{W_d}(q)$ of the polynomial matrix $W_d(q)$ using Algorithm 2.1.

step 9. Construct the designer–assigned row block–matrix W_d using Algorithm 2.2.

step 10. Compute the minimal degree, $\partial[V(q)]_{\min}$, of the denominator, $V(q)$, of the output–feedback compensator (order n_v), such that

$$n_i.\partial[V(q)]_{\min} \geq n_v \, . \tag{4.48}$$

Similarly, compute the minimal degree, $\partial[W_s(q)]_{\min}$, of the solution–assigned closed–loop polynomial matrix, $W_s(q)$ (order n_{w_s}), such that

$$n_i.\partial[W_s(q)]_{\min} \geq n_{w_s} . \tag{4.49}$$

step 11. Set $p_{\min} = \max(\partial[V(q)]_{\min}, \partial[W_s(q)]_{\min})$.

step 12. Apply Algorithm 4.1 to compute the LMB, $\tilde{Y}(q)$, of

$$X(q) = \begin{bmatrix} -W_d(q) \\ \bar{N}_R(q) \\ D_R(q) \end{bmatrix} .$$

Use Gaussian elimination, say, to find the primary dependent rows of the resultant X_R.

step 13. Construct the polynomial matrix $\hat{Y}(q)$. Then proceeding from top to bottom of $\hat{Y}(q)$, we choose those rows of $\hat{Y}(q)$ which correspond to the first n_i linearly independent rows which satisfy Theorem 4.2. This yields the polynomial matrix $Y(q)$ which can be partitioned as

$$Y(q) = \begin{bmatrix} W_s(q) & K(q) & V(q) \end{bmatrix} .$$

step 14. Using Algorithm 2.3, compute the characteristic polynomials $w_s(q)$ and $v(q)$ of the solution–assigned polynomial matrices $W_s(q)$ and $V(q)$, respectively. This calculation yields the values of the solution–assigned closed–loop poles and the compensator poles.

Remark 4.6 Investigating the closed–loop pole positions, while providing stability information, will not suffice in accurately predicting the transient–response characteristics. It is necessary to know the closed–loop zero positions as well as the closed–loop pole positions for each transfer function. Accurate dynamic responses can only then be determined. The technique only provides for pole placement with the closed–loop zero positions constrained by solution. Hence, it may require an iterative process of designer–assigned pole placement in conjunction with the observation of the response of the closed–loop system to achieve a satisfactory transient–response characteristic.

4.3.2.1 Summary

In this section we have considered the problem of solving the MLCE for the design of a proper (causal), left–coprime, least–order dynamic output–feedback compensator, $F(q)$, to realise a proper (causal), stable, closed-loop transfer function, $H(q)$, using arbitrary pole placement. This problem was formulated as a minimal design problem (MDP). We have considered the four conditions necessary for the existence of feasible solutions to the MDP. A method described by Kung and Kailath [57,58,64] for solving the MDP has been adopted and was implemented on a digital computer. This method is numerically more efficient compared to the methods described in [40,80,105,115]. The 'shift–invariance' structure of the resultant X_R can be exploited to derive an algorithm to calculate the left minimal basis, $\tilde{Y}$. The method searches for the primary dependent rows of X_R using, say, Gaussian elimination. This approach, besides being simple, involves operations only on matrices of real numbers rather than on polynomial matrices, and thus is well–suited to computer–aided design.

The design methodology at present cannot assign zeros of the closed–loop system, nor determine the output–feedback compensator out of an infinite set which best satisfies pole assignment and other design requirements simultaneously. An infinite number of feedback compensators exists because the designer–assigned polynomial matrix $W_d(q)$ is *not* unique, given set of designer–assigned closed–loop poles.

4.4 The Pre–compensator

4.4.1 Introduction

The pre–compensator system, $R(q)$, which is shown in Fig. 4.1, is described by the LMFD

$$R(q) = V^{-1}(q).M(q),$$

where the denominator, $V(q)$, is determined from the solution to the output–feedback compensator, $F(q)$; as yet the numerator, $M(q)$, has not been determined. The latter provides in the design process an additional degree of freedom which is used to satisfy steady-state performance specifications.

The steady-state performance of a system may best be described in terms of *errors* when the inputs are simple aperiodic test signals. Two steady-state performance specifications are considered, these being the steady-state *alignment error* and the steady-state *following error*. For SISO systems the term alignment error denotes the steady-state difference between a reference step input of $r(t)$ and a controlled output of $c(t)$. Similarly, the term following error denotes the steady-state difference between a reference ramp input of $r(t)$ and a controlled output of $c(t)$.

In MIMO systems the input and output for which the alignment or following error is to be specified must be explicitly declared. In general, for practical systems the inequality $n_o \geq n_i$ is satisfied. Thus the number of steady-state performance specifications that can be assigned for such a closed-loop system are n_i.

In a MIMO environment the steady-state accuracy is usually not the only steady-state objective under consideration, the level of steady-state decoupling is also of concern. The following methodology takes this additional factor into account in that complete steady-state decoupling is achieved between the controlled outputs for which the steady-state accuracy is specified.

For the purpose of assigning values to the alignment or following error, it is assumed that the system variables are defined in a consistent set of units, or the variables have been normalised, or a per-unit system has been adopted. For example, in a speed control system the reference signal is a voltage and the output speed is given in rad s^{-1}. In this case it may be convenient to convert the reference to a speed signal with the same units, rad s^{-1}.

4.4.2 The Alignment Error

The alignment error is defined by

$$e_a(t) = r(t) - J.c(t) = r(t) - \tilde{c}(t),$$

where:

- $r(t)$ is a $n_i \times 1$ column vector of reference inputs.

- $c(t)$ is a $n_o \times 1$ column vector of outputs.

- $e_a(t)$ denotes a $n_i \times 1$ column vector in which the jth element $e_{a_j}(t)$ represents the alignment error between the jth reference step input $r_j(t)$ and the jth controlled output of $J.c(t)$.

- J denotes a $n_i \times n_o$ matrix of ones and zeros. The purpose of the matrix J is to enable the designer to select the jth reference input of $r(t)$ and the kth output $c(t)$ for which the steady–state error is to be specified, $j \in \underline{n}_i$, $k \in \underline{n}_o$.

- $\tilde{c}(t) = J.c(t)$ is a $n_i \times 1$ vector of controlled outputs.

Let us consider the closed–loop transfer function, eqn. (4.8),

$$H(q) = N_R(q).W^{-1}(q).M(q)\,.$$

Its TFMD is

$$H(q) = \frac{N(q).M(q)}{w(q)}, \tag{4.50}$$

where $N(q) = N_R(q).\mathrm{Adj}(W(q))$ and $w(q) = \det(W(q))$.

Using the TFMD of the closed–loop transfer function, $H(q)$, and the final value theorem, the steady–state alignment error can be derived. For a continuous–time system the error is

$$e_{ac} = E_{ac}.\mathrm{ones}(n_i, 1) = \lim_{s \to 0} s.(I_{n_i} - \frac{J.N(s).M(s)}{w(s)}).r(s),$$

and, in the case of a discrete–time system, the error is

$$e_{ad} = E_{ad}.\mathrm{ones}(n_i, 1) = \lim_{z \to 1} (1 - z^{-1}).(I_{n_i} - \frac{J.N(z).M(z)}{w(z)}).r(z),$$

where:

- e_{ac} and e_{ad} are $n_i \times 1$ column vectors, the elements of which are the continuous– and discrete–time alignment errors, respectively.

- $E_{ac} = \text{diag}[e_{ac}]$ and $E_{ad} = \text{diag}[e_{ad}]$ are $n_i \times n_i$ diagonal matrices, the diagonal elements of which are the alignment errors of the column vectors e_{ac} and e_{ad}, respectively.

- $\text{ones}(n_i, 1)$ is a $n_i \times 1$ column vector of ones.

- $r(s) = K_a.\text{ones}(n_i, 1) \otimes \frac{1}{s}$ and $r(z) = K_a.\text{ones}(n_i, 1) \otimes \frac{z}{z-1}$ are the s– and z–transforms of the step inputs applied to each of the n_i inputs of $r(t)$, respectively.

- K_a is a $n_i \times n_i$ diagonal non–singular matrix, the jth non–zero diagonal element of which corresponds to the magnitude of the jth step input of $r(t)$.

Let us first consider the steady–state alignment error for a continuous–time system,

$$E_{ac}.\text{ones}(n_i, 1) = \lim_{s \to 0} s.(I_{n_i} - \frac{J.N(s).M(s)}{w(s)}).(K_a.\text{ones}(n_i, 1) \otimes \frac{1}{s})$$

or, equivalently,

$$E_{ac} = (I_{n_i} - \frac{J.N(0).M(0)}{w(0)}).K_a \,.$$

To satisfy the steady–state alignment error specifications, the numerator $M(s)$ needs only be a zero degree polynomial matrix; i.e., $M(s) = M_0$. Thus

$$E_{ac} = (I_{n_i} - \frac{J.N(0).M_0}{w(0)}).K_a \,.$$

Rearrangement of the above relationship yields

$$M_0 = M(0) = (J.N(0))^{-1}.w(0).(I_{n_i} - E_{ac}.K_a^{-1}), \tag{4.51}$$

where $w(0) \neq 0$ and $J.N(0)$ is invertible.

Similarly, it can be shown that the relationship for the discrete–time system is

$$M_0 = M(1) = (J.N(1))^{-1}.w(1).(I_{n_i} - E_{ad}.K_a^{-1}), \tag{4.52}$$

where $w(1) \neq 0$ and $J.N(1)$ is invertible. (Note that for those input–output pairs which are to satisfy a finite following–error specification, the associated elements of diag E_{ad} are zero, see § 4.4.4.)

Remark 4.7 Because the error matrices E_{ac} and E_{ad} are diagonal, steady–state decoupling between the controlled outputs used for steady–state alignment error assignment is realised.

4.4.3 The Following Error

The following error is defined by

$$e_f(t) = r(t) - J.c(t) = r(t) - \tilde{c}(t),$$

where:

- $r(t)$ is a $n_i \times 1$ column vector of reference ramp inputs.
- $c(t)$ is a $n_o \times 1$ column vector of outputs.
- $e_f(t)$ denotes a $n_i \times 1$ column vector, in which the jth element $e_{f_j}(t)$ represents the following error between the jth reference ramp input $r_j(t)$ and the jth controlled output of $J.c(t)$.
- $\tilde{c}(t) = J.c(t)$ is a $n_i \times 1$ vector of controlled outputs.

As will be shown shortly, to satisfy a constant following–error specification the alignment error must be zero for a step input. In this case the numerator $M(q)$ must be at least of degree one; i.e.,

$$M(q) = M_0.q + M_1 .$$

Relationships that enable us to calculate the coefficient matrices M_0 and M_1 for a continuous– or discrete–time system will now be derived.

4.4.4 The Following Error for a Continuous–Time System

In the following analysis it is assumed:

- e_{fc} is $n_i \times 1$ column vector, the elements of which are the constant following errors for a continuous–time system.
- $E_{fc} = \text{diag}[e_{fc}]$ is a $n_i \times n_i$ diagonal matrix, the diagonal elements of which are the following errors of the column vector e_{fc}.
- $r(s) = K_f.\text{ones}(n_i, 1) \otimes 1/s^2$ is the s transform of the n_i ramp inputs of $r(t)$.

- K_f is a $n_i \times n_i$ diagonal non-singular matrix, the jth non-zero diagonal element of which corresponds to the slope of the jth ramp input of $r(t)$.

It is shown in the following theorem that to satisfy a constant following-error specification the alignment error must be zero. In order to simplify this theorem, it is assumed that the i th input and the i th output form an input/controlled-output pair for specifying the errors (e.g., J is the identity matrix). The theorem is valid for any J matrix defined as in § 4.4.2.

The closed-loop transfer function $H(s)$ is given by eqn. (4.50). Let

- $H_{ii}(s)$ be the transfer function between the ith input-output pair, and
- $1 - H_{ii}(s) = s\widehat{H}_{ii}(s)$.

Theorem 4.3 Assume, for the i th input-output pair of the closed-loop system $H(s)$, the following error is constant, i.e.,

$$e_{fc\,i} = \lim_{t\to\infty} (r_i(t) - c_i(t)) < \infty ,$$

for the ramp input

$$r_i(t) = K_{f\,i,i} \,.\, t\,, \quad t > 0\,.$$

Then the alignment error is zero, i.e.

$$e_{ac\,i} = \lim_{t\to\infty} (r_i(t) - c_i(t)) = 0\,,$$

for the step input $r_i(t) = K_{a\,i,i}\,,\quad t > 0$, if $\lim_{s\to 0} \widehat{H}_{ii}(s) < \infty$.

Proof 4.2 The response of the i th output to a ramp input $r_i(s) = K_{f\,i,i}\,/s^2$ is $c_i(s) = H_{ii}(s)\,r_i(s)$. Then the following error

$$e_{fc\,i} = \lim_{s\to 0} s(r_i(s) - c_i(s)) = \lim_{s\to 0} s(1 - H_{ii}(s))\,.\,K_{f\,i,i}\,/s^2 < \infty$$

if

$$1 - H_{ii}(s) = s\widehat{H}_{ii}(s) \text{ and } \lim_{s\to 0} \widehat{H}_{ii}(s) < \infty\,.$$

Thus, for a step input,

$$\lim_{s\to 0} s(r_i(s) - c_i(s)) = \lim_{s\to 0} s(1 - H_{ii}(s)) \,.\, K_{a\,i,i} /s = \lim_{s\to 0} s\widehat{H}_{ii}(s) \,.\, K_{a\,i,i} = 0 \,.$$

∇∇∇

Remark 4.8 Assume

$$H_{ii}(s) = \frac{b_m s^m + \ldots + b_0}{s^n + a_{n-1}s^{n-1} + \ldots + a_0},$$

for $m \le n$. Then

$$s\widehat{H}_{ii}(s) = 1 - H_{ii}(s) = \frac{s^n + \ldots + (a_m - b_m)s^m + \ldots + (a_0 - b_0)}{s^n + a_{n-1}s^{n-1} + \ldots + a_0},$$

or,

$$\widehat{H}_{ii}(s) = \frac{s^n + \ldots + (a_m - b_m)s^m + \ldots + (a_0 - b_0)}{s\,(s^n + a_{n-1}s^{n-1} + \ldots + a_0)}.$$

This result implies that $a_0 = b_0$ to satisfy the condition $\lim_{s\to 0} \widehat{H}_{ii}(s) < \infty$ for constant following error. Because a_0 and b_0 may each be some function of the plant parameters they, and thus the accuracy of the control system, may be sensitive to parameter variations. Note, from the above proof, that the condition for the following error is to be zero is that $a_1 = b_1$.

∇∇∇

Let us first satisfy the specifications on the alignment errors for the n_i input–output pairs $r_i(t)$, $\tilde{c}_i(t)$. If a following error has been specified for such a pair, the alignment error is set to zero. Equation (4.51) is used to solve for M_1, i.e.

$$M_1 = M(0) = (J.N(0))^{-1}.w(0).(I_{n_i} - E_{ac}.K_a^{-1}). \tag{4.53}$$

The steady–state following error is

$$e_{fc} = E_{fc}.\text{ones}(n_i, 1) = \lim_{s\to 0} s.(I_{n_i} - \frac{J.N(s).M(s)}{w(s)}).K_f.\text{ones}(n_i, 1) \otimes \frac{1}{s^2}. \tag{4.54}$$

The limit in eqn. (4.54) is indeterminate; application of L'Hôpital's rule yields

$$E_{fc} = -J.\frac{[\dot{N}(0).M(0) + N(0).\dot{M}(0)].w(0) - N(0).M(0).\dot{w}(0)}{w^2(0)}.K_f,$$

where $\cdot = \frac{d}{ds}$. Rearranging the above relationship we find

$$M_0 = \frac{(J.N(0))^{-1}}{w(0)}.(-E_{\rm fc}.K_{\rm f}^{-1}.w^2(0) + J.[N(0).\dot{w}(0) - \dot{N}(0).w(0)].M_1), \qquad (4.55)$$

where $M_1 = M(0)$ and $M_0 = \dot{M}(0)$.

Remark 4.9 There is a problem if the steady–state performance requirements contain a mix of alignment– and following–error specifications. (For each input–output pair either an alignment error or a following error is specified.) An examination of eqn. (4.55) reveals that, for an input–output pair which is required to satisfy only an alignment–error specification (which may *not* be zero), a 'following error' must also be specified. This has two disadvantages. Firstly, deciding on the value of the following error may be a trial–and–error process. Secondly, in attempting to accommodate the 'non–specified' as well as the specified following errors, it may not be possible to realize a proper compensator. In the examples discussed in Chap. 5 it will be assumed the steady–state performance specifications are given in terms of either the alignment errors or the following errors, but not both.

4.4.5 The Following Error for a Discrete–Time System

A similar procedure to that used for the continuous–time case is employed to derive the relationships for the discrete–time system.

The specifications on the alignment errors for the $n_{\rm i}$ input–output pairs $r_i(t)$, $\tilde{c}_i(t)$, are satisfied first. If a following error has been specified for such a pair, the associated alignment error is set to zero. Equation (4.52) is used to solve for $M(1)$, i.e.

$$M(1) = M_0 + M_1 = (J.N(1))^{-1}.w(1).(I_{n_{\rm i}} - E_{\rm ad}.K_{\rm a}^{-1})\,. \qquad (4.56)$$

The steady–state following error is

$$e_{\rm fd} = E_{\rm fd}.\text{ones}(n_{\rm i},1) = \lim_{z\to 1}(1-z^{-1}).(I_{n_{\rm i}} - \frac{J.N(z).M(z)}{w(z)}).K_{\rm f}.\text{ones}(n_{\rm i},1) \otimes \frac{T.z}{(z-1)^2},$$

where:

- T is the sampling period.

- e_{fd} is $n_{\mathrm{i}} \times 1$ column vector, the elements of which are the constant following errors of a discrete–time system.

- $E_{\mathrm{fd}} = \mathrm{diag}[e_{\mathrm{fd}}]$ is a $n_{\mathrm{i}} \times n_{\mathrm{i}}$ diagonal matrix, the diagonal elements of which are the following errors of the column vector e_{fd}.

- $r(z) = K_{\mathrm{f}}.\mathrm{ones}(n_{\mathrm{i}}, 1) \otimes T.z/(z-1)^2$ is the z transform of the n_{i} ramp inputs of $r(t)$.

- K_{f} is a $n_{\mathrm{i}} \times n_{\mathrm{i}}$ diagonal non–singular matrix, the ith non–zero diagonal element of which corresponds to the slope of the ith ramp input of $r(t)$.

Appling L'Hôpital's rule, we find

$$E_{\mathrm{fd}} = -T.J.\frac{[\dot{N}(1).M(1) + N(1).\dot{M}(1)].w(1) - N(1).M(1).\dot{w}(1)}{w^2(1)}.K_{\mathrm{f}},$$

where $\cdot = \frac{d}{dz}$, $M(1) = M_0 + M_1$ and $\dot{M}(1) = M_0$. Thus

$$E_{\mathrm{fd}} = -T.J.\frac{[\dot{N}(1).(M_0 + M_1) + N(1).M_0].w(1) - N(1).(M_0 + M_1)\dot{w}(1)}{w^2(1)}.K_{\mathrm{f}}\,.$$

Hence

$$J.[(\dot{N}(1) + N(1)).w(1) - \dot{w}(1).N(1)].M_0 + J.[\dot{N}(1).w(1) - N(1).\dot{w}(1)].M_1$$
$$= -E_{\mathrm{fd}}.K_g^{-1}.\frac{w^2(1)}{T} \qquad (4.57)$$

Combining eqns. (4.56) and (4.57) a set of linear equations described by eqn. (4.58) is constructed. Thus the unknown coefficient block–matrices M_0 and M_1 of $M(z)$ can be determined such that the requirements for zero alignment error and constant following error are satisfied.

$$\begin{bmatrix} I_{n_{\mathrm{i}}} & I_{n_{\mathrm{i}}} \\ J.[(\dot{N}(1) + N(1)).w(1) - \dot{w}(1).N(1)] & J.[\dot{N}(1).w(1) - N(1).\dot{w}(1)] \end{bmatrix} \cdot \begin{bmatrix} M_0 \\ M_1 \end{bmatrix} \qquad (4.58)$$
$$= \begin{bmatrix} (J.N(1))^{-1}.w(1).(I_{n_{\mathrm{i}}} - E_{\mathrm{ad}}.K_{\mathrm{a}}^{-1}) \\ -E_{\mathrm{fd}}.K_g^{-1}.\frac{w^2(1)}{T} \end{bmatrix}$$

Comment: Remark 4.9 applies to the discrete–time as well. A solution for the numerator $M(q)$ for a following error specification does not ensure that the resulting pre–compensator, $R(q)$, is causal. The condition $\partial_{rj}[M(q)] \leq \partial_{rj}[V(q)]$, $j \in \underline{n_{\mathrm{i}}}$, must be satisfied.

Remark 4.10 Steady–state decoupling between the controlled outputs used for following–error assignment is realised when the error matrices E_{fc} and E_{fd} are diagonal.

4.4.6 Summary

The design of a continuous– or discrete–time MIMO pre–compensator, $R(q) = V^{-1}(q).M(q)$, to satisfy the steady–state specifications has been discussed in this section. The parameters of the numerator $M(q)$ have been determined such that the requirements on the steady–state alignment or following error are met. The formulation of the problem permits the decoupling of the errors to occur in the steady state. A mix of error specifications may be provided, i.e. for each input–output pair either an alignment error or a following error is specified. However, the design procedure is complicated by such a mix.

4.5 Example of the Design Calculations

To illustrate the results of this chapter let us consider the following example.

Example 4.3 Consider the controllable and observable MIMO continuous–time plant described by the equations

$$\dot{x}(t) = \begin{bmatrix} -1 & 0 & 3 & -1 \\ 4 & -2 & 1 & 2 \\ 3 & 0 & -3 & 4 \\ -4 & -2 & -3 & -4 \end{bmatrix} .x(t) + \begin{bmatrix} 2 & 1 \\ 0 & 2 \\ 1 & -1 \\ -1 & 0 \end{bmatrix} .u(t)$$

$$c(t) = \begin{bmatrix} -1 & 2 & 1 & -2 \\ 0 & -1 & -2 & 0 \\ 1 & 1 & 0 & -2 \end{bmatrix} .x(t)$$

with $n_{\mathrm{s}} = 4$ states, $n_{\mathrm{i}} = 2$ actuating inputs and $n_{\mathrm{o}} = 3$ controlled outputs. The poles of the plant are

$$-0.8438 \pm j1.81521 \quad \text{and} \quad -4.1562 \pm j1.6405\ .$$

A continuous–time dynamic output–feedback compensator, $F(s)$, and a continuous–time pre–compensator, $R(s)$, are to be designed for this plant such that the pole locations and the steady–state performance of the resulting closed–loop system satisfy the following requirements.

The output–feedback compensator is to be designed such that the four plant poles are shifted to the positions:

$$-3 \pm j4.5 \text{ and } \quad -5 \pm j2$$

in the s–plane.

The pre–compensator, $R(s)$, is to be designed such that the closed-loop system satisfies the following steady–state performance specifications:

1. The steady–state alignment error at the controlled output $c_3(t)$ when a unit step, $U(t)$, is applied at the reference input $r_1(t)$ is to be zero.

2. The steady–state alignment error at the controlled output $c_1(t)$ for a step input of $2.5U(t)$ applied at the reference input $r_2(t)$ is to be 0.8.

The computations are carried out with the aid of a software package using double precision arithmetic.

Dynamic performance considerations

Transforming the SSD of the plant to its equivalent RMFD, using Algorithm 3.1, the denominator and numerator matrices are given by

$$D_R = \left[\begin{array}{cc|cc|cc} 1.0000 & 0.0000 & -3.8175 & -7.1314 & 2.5109 & 0.8321 \\ 0.0000 & 1.0000 & 6.6423 & 13.8175 & 27.1825 & 40.8686 \end{array}\right]$$

or, equivalently,

$$D_R(s) = \begin{bmatrix} s^2 - 3.8175s + 2.5109 & -7.1314s + 0.8321 \\ 6.6423s + 27.1825 & s^2 + 13.8175s + 40.8686 \end{bmatrix}$$

having degree $\partial[D_R(s)] = 2$, and

$$N_R = \left[\begin{array}{cc|cc} 1.0000 & 2.0000 & 34.4672 & 38.5036 \\ -2.0000 & 0.0000 & 2.6350 & 3.2628 \\ 4.0000 & 3.0000 & 27.6569 & 17.9270 \end{array}\right]$$

or, equivalently,

$$N_R(s) = \begin{bmatrix} s + 34.4672 & 2s + 38.5036 \\ -2s + 2.6350 & 3.2628 \\ 4s + 27.6569 & 3s + 17.9270 \end{bmatrix}$$

having degree $\partial[N_R(s)] = 1$.

The controlled outputs $c_1(t)$ and $c_3(t)$ of the plant are available for output feedback. Thus $n_f = 2$ and the feedback interconnection matrix is

$$U = \begin{bmatrix} 1 & 0 & 0 \\ 0 & 0 & 1 \end{bmatrix}.$$

The numerator of the modified plant is

$$\bar{N}_R = U.N_R = \left[\begin{array}{cc|cc} 1.0000 & 2.0000 & 34.4672 & 38.5036 \\ 4.0000 & 3.0000 & 27.6569 & 17.9270 \end{array} \right]$$

or, equivalently,

$$\bar{N}_R(s) = \begin{bmatrix} s + 34.4672 & 2s + 38.5036 \\ 4s + 27.6569 & 3s + 17.9270 \end{bmatrix}$$

having degree $\partial[\bar{N}_R(s)] = 1$. The modified plant $G_F(s) = \bar{N}_R(s).D_R^{-1}(s)$ is right–coprime.

Let us determine the minimal order, output–feedback compensator which enables us to arbitrarily assign no less than four closed–loop system poles. We begin with a zero–order feedback compensator and increment the order, if necessary, until an acceptable number of designer–assigned and solution–assigned closed–loop poles results.

A static output–feedback compensator allows no greater than

$$n_{w_d} = \min(n_s = 4, (n_f = 2) + (n_i = 2) - 1) = 3$$

closed–loop poles to be designer–assigned. This is unsatisfactory. To satisfy the closed–loop specifications, the order of the output feedback is incremented to $n_v = 1$. A first–order output–feedback compensator allows no greater than

$$n_{w_d} = \min((n_s = 4) + (n_v = 1), (n_f = 2) + (n_i = 2) - 1 + (2n_v = 2)) = 5$$

closed–loop poles to be designer–assigned. This is an acceptable configuration. Thus, an output–feedback compensator of order $n_v = 1$, at least, is required to assign four closed–loop poles. The five closed–loop poles are chosen to be designer–assigned poles.

The extra closed–loop pole is chosen to be at -10 in order that it has minimal effect on the response of the closed–loop system.

The standard form of the column structure matrix of the designer–assigned, closed–loop polynomial matrix $W_d(s)$ of order $n_{w_d} = 5$ is given by

$$P_{W_d}(s) = \begin{bmatrix} 0 & s^3 \\ s^2 & 0 \end{bmatrix},$$

where the degree of $W_d(s)$ is $\partial[W_d(s)] = 3$.

The designer–assigned closed–loop poles are now:

$$-3 \pm j4.5, \quad -5 \pm j2 \text{ and } \quad -10 .$$

Applying Algorithm 2.2 the row block–matrix description, W_d, of the designer–assigned polynomial matrix $W_d(s)$ is determined,

$$W_d = \left[\begin{array}{cc|cc|cc} 0.0000 & 1.0000 & 0.0000 & 20.0000 & 0.0000 & 129.0000 \\ 0.0000 & 0.0000 & 1.0000 & 0.0000 & 6.0000 & 0.0000 \end{array} \cdots \right.$$

$$\left. \cdots \begin{array}{|cc} 0.0000 & 290.0000 \\ 29.2500 & 0.0000 \end{array} \right]$$

or, equivalently,

$$W_d(s) = \begin{bmatrix} 0 & s^3 + 20s^2 + 129s + 290 \\ s^2 + 6s + 29.25 & 0 \end{bmatrix} . \tag{4.59}$$

Let us now determine the lower bound on the degree p of $\tilde{Y}(s)$. Applying the inequalities of eqn. (4.48) and (4.49) the minimal degree for $W_s(s)$ and $V(s)$ is found to be $\partial[W_s(s)]_{\min} = 0$ and $\partial[V(s)]_{\min} = 1$, respectively. Thus

$$p_{\min} = \max((\partial[W_s(s)]_{\min} = 0), (\partial[V(s)] = 1)) = 1 .$$

The polynomial matrix $X(s)$ and its row block matrix X for this example are

$$X(s) = \begin{bmatrix} -W_d(s) \\ \tilde{N}_R(s) \\ D_R(s) \end{bmatrix}$$

$$= \begin{bmatrix} 0 & -s^3 - 20s^2 - 129s - 290 \\ -s^2 - 6s - 29.25 & 0 \\ s + 34.4672 & 2s + 38.5036 \\ 4s + 27.6569 & 3s + 17.9270 \\ s^2 - 3.8175s + 2.5109 & -7.1314s + 0.8321 \\ 6.6423s + 27.1825 & s^2 + 13.8175s + 40.8686 \end{bmatrix}$$

and

$$X = \left[\begin{array}{cc|cc|cc} 0.0000 & -1.0000 & 0.0000 & -20.0000 & 0.0000 & -129.0000 \\ 0.0000 & 0.0000 & -1.0000 & 0.0000 & -6.0000 & 0.0000 \\ 0.0000 & 0.0000 & 0.0000 & 0.0000 & 1.0000 & 2.0000 \\ 0.0000 & 0.0000 & 0.0000 & 0.0000 & 4.0000 & 3.0000 \\ 0.0000 & 0.0000 & 1.0000 & 0.0000 & -3.8175 & -7.1314 \\ 0.0000 & 0.0000 & 0.0000 & 1.0000 & 6.6423 & 13.8175 \end{array}\right. \cdots$$

$$\cdots \left.\begin{array}{|cc} 0.0000 & -290.0000 \\ -29.2500 & 0.0000 \\ 34.4672 & 38.5036 \\ 27.6569 & 17.9270 \\ 2.5109 & 0.8321 \\ 27.1825 & 40.8686 \end{array}\right]$$

with $\bar{n} = (n_{\rm f} + 2n_{\rm i}) = 6$, $n = \bar{n} - n_{\rm i} = 4$.

Note that the dimension of the left null space of $X(s)$ is $n = 4$, i.e. $\tilde{Y}(s)$ is a 4×6 polynomial matrix having rank 4. Therefore, there are only four primary dependent rows in the corresponding matrix $\mathcal{X}(s)$.

It is found that the degree p of $\tilde{Y}$ (or $\tilde{Y}(s)$) required to enable $n = 4$ primary dependent rows of X_R (or $\mathcal{X}(s)$) to be found is 2. Note, however, the degree p of $\tilde{Y}$ would, generally, be determined in the course of the computation. The matrices $\mathcal{X}(s)$ and X_{R} are given by

$$\mathcal{X}(s) = \begin{bmatrix} 0 & -s^5 - 20s^4 - 129s^3 - 290s^2 \\ -s^4 - 6s^3 - 29.25s^2 & 0 \\ s^3 + 34.4672s^2 & 2s^3 + 38.5036s^2 \\ 4s^3 + 27.6569s^2 & 3s^3 + 17.9270s^2 \\ s^4 - 3.8175s^3 + 2.5109s^2 & -7.1314s^3 + 0.8321s^2 \\ 6.6423s^3 + 27.1825s^2 & s^4 + 13.8175s^3 + 40.8686s^2 \\ -- & -- \\ 0 & -s^4 - 20s^3 - 129s^2 - 290s \\ -s^3 - 6s^2 - 29.25s & 0 \\ s^2 + 34.4672s & 2s^2 + 38.5036s \\ 4s^2 + 27.6569s & 3s^2 + 17.9270s \\ s^3 - 3.8175s^2 + 2.5109s & -7.1314s^2 + 0.8321s \\ 6.6423s^2 + 27.1825s & s^3 + 13.8175s^2 + 40.8686s \\ -- & -- \\ 0 & -s^3 - 20s^2 - 129s - 290 \\ -s^2 - 6s - 29.25 & 0 \\ s + 34.4672 & 2s + 38.5036 \\ 4s + 27.6569 & 3s + 17.9270 \\ s^2 - 3.8175s + 2.5109 & -7.1314s + 0.8321 \\ 6.6423s + 27.1825 & s^2 + 13.8175s + 40.8686 \end{bmatrix}$$

and

$$X_{\mathrm{R}} = \langle X \rangle_{(p=2)} = \begin{bmatrix} 0.0000 & -1.0000 & 0.0000 & -20.0000 & 0.0000 \\ 0.0000 & 0.0000 & -1.0000 & 0.0000 & -6.0000 \\ 0.0000 & 0.0000 & 0.0000 & 0.0000 & 1.0000 \\ 0.0000 & 0.0000 & 0.0000 & 0.0000 & 4.0000 \\ 0.0000 & 0.0000 & 1.0000 & 0.0000 & -3.8175 \\ 0.0000 & 0.0000 & 0.0000 & 1.0000 & 6.6423 \\ 0.0000 & 0.0000 & 0.0000 & -1.0000 & 0.0000 \\ 0.0000 & 0.0000 & 0.0000 & 0.0000 & -1.0000 \\ 0.0000 & 0.0000 & 0.0000 & 0.0000 & 0.0000 \\ 0.0000 & 0.0000 & 0.0000 & 0.0000 & 0.0000 \\ 0.0000 & 0.0000 & 0.0000 & 0.0000 & 1.0000 \\ 0.0000 & 0.0000 & 0.0000 & 0.0000 & 0.0000 \\ 0.0000 & 0.0000 & 0.0000 & 0.0000 & 0.0000 \\ 0.0000 & 0.0000 & 0.0000 & 0.0000 & 0.0000 \\ 0.0000 & 0.0000 & 0.0000 & 0.0000 & 0.0000 \\ 0.0000 & 0.0000 & 0.0000 & 0.0000 & 0.0000 \\ 0.0000 & 0.0000 & 0.0000 & 0.0000 & 0.0000 \\ 0.0000 & 0.0000 & 0.0000 & 0.0000 & 0.0000 \end{bmatrix} \cdots$$

$$
\cdots \left. \begin{matrix}
-129.00 & 0.0000 & -290.00 & 0.0000 & 0.0000 & 0.0000 & 0.0000 \\
0.0000 & -29.250 & 0.0000 & 0.0000 & 0.0000 & 0.0000 & 0.0000 \\
2.0000 & 34.467 & 38.504 & 0.0000 & 0.0000 & 0.0000 & 0.0000 \\
3.0000 & 27.657 & 17.927 & 0.0000 & 0.0000 & 0.0000 & 0.0000 \\
-7.1314 & 2.5109 & 0.8321 & 0.0000 & 0.0000 & 0.0000 & 0.0000 \\
13.817 & 27.182 & 40.869 & 0.0000 & 0.0000 & 0.0000 & 0.0000 \\
-20.000 & 0.0000 & -129.00 & 0.0000 & -290.00 & 0.0000 & 0.0000 \\
0.0000 & -6.0000 & 0.0000 & -29.250 & 0.0000 & 0.0000 & 0.0000 \\
0.0000 & 1.0000 & 2.0000 & 34.467 & 38.504 & 0.0000 & 0.0000 \\
0.0000 & 4.0000 & 3.0000 & 27.657 & 17.927 & 0.0000 & 0.0000 \\
0.0000 & -3.8175 & -7.1314 & 2.5109 & 0.8321 & 0.0000 & 0.0000 \\
1.0000 & 6.6423 & 13.817 & 27.182 & 40.869 & 0.0000 & 0.0000 \\
-1.0000 & 0.0000 & -20.000 & 0.0000 & -129.00 & 0.0000 & -290.00 \\
0.0000 & -1.0000 & 0.0000 & -6.0000 & 0.0000 & -29.250 & 0.0000 \\
0.0000 & 0.0000 & 0.0000 & 1.0000 & 2.0000 & 34.467 & 38.504 \\
0.0000 & 0.0000 & 0.0000 & 4.0000 & 3.0000 & 27.657 & 17.927 \\
0.0000 & 1.0000 & 0.0000 & -3.8175 & -7.1314 & 2.5109 & 0.8321 \\
0.0000 & 0.0000 & 1.0000 & 6.6423 & 13.817 & 27.182 & 40.869
\end{matrix} \right] .
$$

We now apply Algorithm 4.1 to find the primary dependent rows of X_{R} and the corresponding dependence relationships. The results are as follows:

$i = 1$: $\eta_1 = 9$. As shown in eqn. (4.44), the linear dependence can be expressed as,

$$
\begin{aligned}
X_{\mathrm{R}\,9} = 0.5196 X_{\mathrm{R}\,7} + 0.5196 X_{\mathrm{R}\,6} - 0.4297 X_{\mathrm{R}\,5} + 0.8970 X_{\mathrm{R}\,4} - 7.6891 X_{\mathrm{R}\,3} \\
+0.1187 X_{\mathrm{R}\,2} + 6.2127 X_{\mathrm{R}\,1}.
\end{aligned}
$$

Therefore, $\tilde{Y}_{1,9} = 1$, $\tilde{Y}_{1,7} = -0.5196$, $\tilde{Y}_{1,6} = -0.5196$, $\tilde{Y}_{1,5} = 0.4297$, $\tilde{Y}_{1,4} = -0.8970$, $\tilde{Y}_{1,3} = 7.6891$, $\tilde{Y}_{1,2} = -0.1187$, $\tilde{Y}_{1,1} = -6.2127$.

Eliminate row 9 and rows $\eta_1 \bmod \bar{n}$, i.e. row 15, from X_R.

$i = 2$: $\eta_2 = 10$,

$$
\begin{aligned}
X_{\mathrm{R}\,10} = 0.3338 X_{\mathrm{R}\,7} + 0.3338 X_{\mathrm{R}\,6} - 6.5386 X_{\mathrm{R}\,5} + 7.4322 X_{\mathrm{R}\,4} - 19.4674 X_{\mathrm{R}\,3} \\
-7.7557 X_{\mathrm{R}\,2} + 4.0636 X_{\mathrm{R}\,1}.
\end{aligned}
$$

Therefore, $\tilde{Y}_{2,10} = 1$, $\tilde{Y}_{2,7} = -0.3338$, $\tilde{Y}_{2,6} = -0.3338$, $\tilde{Y}_{2,5} = 6.5386$, $\tilde{Y}_{2,4} = -7.4322$, $\tilde{Y}_{2,3} = 19.4674$, $\tilde{Y}_{2,2} = 7.7557$, $\tilde{Y}_{2,1} = -4.0636$.

Eliminate row 10 and rows η_2 mod $\bar{n}$, i.e. row 16.

$i = 3$: $\eta_3 = 11$,

$$X_{\mathrm{R}\,11} = -X_{\mathrm{R}\,8} - 0.8470X_{\mathrm{R}\,7} - 0.8470X_{\mathrm{R}\,6} - 14.8140X_{\mathrm{R}\,5} + 19.0359X_{\mathrm{R}\,4}$$

$$-25.5093X_{\mathrm{R}\,3} - 19.0055X_{\mathrm{R}\,2} - 12.3679X_{\mathrm{R}\,1}.$$

Therefore, $\tilde{Y}_{3,11} = 1$, $\tilde{Y}_{3,8} = 1$, $\tilde{Y}_{3,7} = 0.8470$, $\tilde{Y}_{3,6} = 0.8470$, $\tilde{Y}_{3,5} = 14.8140$, $\tilde{Y}_{3,4} = -19.0359$, $\tilde{Y}_{3,3} = 25.5093$, $\tilde{Y}_{3,2} = 19.0055$, $\tilde{Y}_{3,1} = 12.3679$.

Eliminate row 11 and rows η_3 mod $\bar{n}$, i.e. row 17.

$i = 4$: $\eta_4 = 13$,

$$X_{\mathrm{R}\,13} = -X_{\mathrm{R}\,12} + 6.6423X_{\mathrm{R}\,8} - 1.1116X_{\mathrm{R}\,7} + 5.0708X_{\mathrm{R}\,6} + 2.3248X_{\mathrm{R}\,5}$$

$$-18.0229X_{\mathrm{R}\,4} - 55.4425X_{\mathrm{R}\,3} + 62.2484X_{\mathrm{R}\,2} + 76.0147X_{\mathrm{R}\,1}.$$

Therefore, $\tilde{Y}_{4,13} = 1$, $\tilde{Y}_{4,12} = 1$, $\tilde{Y}_{4,8} = -6.6423$, $\tilde{Y}_{4,7} = 1.1116$, $\tilde{Y}_{4,6} = -5.0708$, $\tilde{Y}_{4,5} = -2.3248$, $\tilde{Y}_{4,4} = 18.0229$, $\tilde{Y}_{4,3} = 55.4425$, $\tilde{Y}_{4,2} = -62.2484$, $\tilde{Y}_{4,1} = -76.0147$.

Since $i = 4 = n$, the algorithm terminates. The row block–matrix of the LMB $\tilde{Y}$ is given by

$$\tilde{Y} = \left[\begin{array}{cccccc} 0.0000 & 0.0000 & 0.0000 & 0.0000 & 0.0000 & 0.0000 \\ 0.0000 & 0.0000 & 0.0000 & 0.0000 & 0.0000 & 0.0000 \\ 0.0000 & 0.0000 & 0.0000 & 0.0000 & 0.0000 & 0.0000 \\ 0.0000 & 0.0000 & 0.0000 & 0.0000 & 0.0000 & 1.0000 \end{array} \right. \cdots$$

$$\cdots \left| \begin{array}{cccccc} 0.0000 & 0.0000 & 0.0000 & 1.0000 & 0.0000 & -0.5196 \\ 0.0000 & 0.0000 & 1.0000 & 0.0000 & 0.0000 & -0.3338 \\ 0.0000 & 1.0000 & 0.0000 & 0.0000 & 1.0000 & 0.8470 \\ 1.0000 & 0.0000 & 0.0000 & 0.0000 & -6.6423 & 1.1116 \end{array} \right| \cdots$$

$$\cdots \left. \begin{array}{cccccc} -0.5196 & 0.4297 & -0.8970 & 7.6891 & -0.1187 & -6.2127 \\ -0.3338 & 6.5386 & -7.4322 & 19.4674 & 7.7557 & -4.0636 \\ 0.8470 & 14.8140 & -19.0359 & 25.5093 & 19.0055 & 12.3679 \\ -5.0708 & -2.3248 & 18.0229 & 55.4425 & -62.2484 & -76.0147 \end{array} \right]$$

or, equivalently,

$$\tilde{Y}(s) = \begin{bmatrix} -0.5196 & 0.4297 & -0.8970 & \underline{s+7.6891} & -0.1187 \\ -0.3338 & 6.5386 & \underline{s-7.4322} & 19.4674 & 7.7557 \\ 0.8470 & \underline{s+14.8140} & -19.0359 & 25.5093 & s+19.0055 \\ s-5.0708 & -2.3248 & 18.0229 & 55.4425 & -6.6423s-62.2484 \end{bmatrix} \cdots$$

$$\cdots \begin{matrix} -0.5196s-6.2127 \\ -0.3338s-4.0636 \\ 0.8470s+12.3679 \\ \underline{s^2+1.1116s-76.0147} \end{matrix} \Bigg] .$$

The dimension of $\tilde{Y}(s)$ is $n = (n_{\rm i} + n_{\rm f}) = 4$ by $\bar{n} = (n_{\rm f} + 2n_{\rm i}) = 6$, the row indices are $\mu_1 = 1$, $\mu_2 = 1$, $\mu_3 = 1$ and $\mu_4 = 2$, and the pivot indices are $\rho_1 = 3$, $\rho_2 = 4$, $\rho_3 = 5$ and $\rho_4 = 1$ (the corresponding elements of $\tilde{Y}(s)$ are underlined). It can be verified that $\tilde{Y}(s)$ satisfies the conditions of Definition 4.2. It is therefore in Popov echelon form. It should be noted that $\tilde{Y}(s)$ is uniquely defined, i.e., it is in canonical form.

In extracting the polynomial matrix $\hat{C}(s) = [\widehat{K}(s)\ \hat{V}(s)]$ from $\tilde{Y}(s)$ it is found that the submatrix $[\tilde{K}(s)\,\tilde{V}(s)]$ is a full row rank, row-proper polynomial matrix, and thus, for this example $\hat{C}(s) = [\tilde{K}(s)\ \tilde{V}(s)] = [\widehat{K}(s)\ \hat{V}(s)]$ and $\hat{n} = n = 4$. The highest row-degree coefficient matrix, $\Gamma_{\rm r}[\hat{C}(s)]$, of $\hat{C}(s)$ is described by

$$\Gamma_{\rm r}[\hat{C}(s)] = [\widehat{K}_\Gamma\ \hat{V}_\Gamma]$$

$$= \left[\begin{array}{cc|cc} 0.0000 & 1.0000 & 0.0000 & -0.5196 \\ 1.0000 & 0.0000 & 0.0000 & -0.3338 \\ 0.0000 & 0.0000 & 1.0000 & 0.8470 \\ 0.0000 & 0.0000 & 0.0000 & 1.0000 \end{array}\right]$$

which is of full row rank, and thus $\hat{C}(s)$ is row proper. Note for this example $\hat{Y}(s) = \tilde{Y}(s)$, since $[\tilde{K}(s)\ \tilde{V}(s)]$ is a full row rank, row-proper matrix.

Proceeding from top to bottom of $\hat{Y}(s) = \tilde{Y}(s)$, we choose those rows of $\hat{Y}(s)$ which correspond to the first $n_{\rm i} = 2$ linearly independent rows of $\hat{V}_\Gamma$ (see Theorem 4.2). These being rows $\{1, 3\}$ of $\hat{Y}(s)$. This yields the row block-matrix

$$Y = \begin{bmatrix} 0.0000 & 0.0000 & 0.0000 & 0.0000 & 0.0000 & 0.0000 \\ 0.0000 & 0.0000 & 0.0000 & 0.0000 & 0.0000 & 0.0000 \end{bmatrix} \cdots$$

$$\cdots \left| \begin{matrix} 0.0000 & 0.0000 & 0.0000 & 1.0000 & 0.0000 & -0.5196 \\ 0.0000 & 1.0000 & 0.0000 & 0.0000 & 1.0000 & 0.8470 \end{matrix} \right| \cdots$$

$$\cdots \begin{matrix} -0.5196 & 0.4297 & -0.8970 & 7.6891 & -0.1187 & -6.2127 \\ 0.8470 & 14.8140 & -19.0359 & 25.5093 & 19.0055 & 12.3679 \end{matrix} \Bigg]$$

or, equivalently,

$$Y(s) = \begin{bmatrix} -0.5196 & 0.4297 & -0.8970 & s + 7.6891 & -0.1187 \\ 0.8470 & s + 14.8140 & -19.0359 & 25.5093 & s + 19.0055 \end{bmatrix} \cdots$$

$$\cdots \begin{matrix} -0.5196s - 6.2127 \\ 0.8470s + 12.3679 \end{matrix} \Bigg]$$

of dimension 2×6. Note that rows $\{2, 3\}$ also yield a proper (causal), coprime and least–order solution. However, in the course of the calculations it is found that, of the acceptable solutions, an output–feedback compensator of order $n_v = 2$ is required to assign five closed–loop poles exactly, rather than an order of $n_v = 1$. Thus the order of solution–assigned closed–loop pole is one and the order of the closed–loop system is $n_w = 6$.

Extracting the row block–matrices W_s, K and V from Y we have, firstly,

$$W_s = \left[\begin{array}{cc|cc} 0.0000 & 0.0000 & -0.5196 & 0.4297 \\ 0.0000 & 1.0000 & 0.8470 & 14.8140 \end{array} \right]$$

or, equivalently,

$$W_s(s) = \begin{bmatrix} -0.5196 & 0.4297 \\ 0.8470 & s + 14.8140 \end{bmatrix}$$

having dimension 2×2 and degree $\partial[W_s(s)] = 1$. Secondly,

$$K = \left[\begin{array}{cc|cc} 0.0000 & 1.0000 & -0.8970 & 7.6891 \\ 0.0000 & 0.0000 & -19.0359 & 25.5093 \end{array} \right]$$

or, equivalently,

$$K(s) = \begin{bmatrix} -0.8970 & s + 7.6891 \\ -19.0359 & 25.5093 \end{bmatrix}$$

having dimension 2×2 and degree $\partial[K(s)] = 1$. Finally,

$$V = \left[\begin{array}{cc|cc} 0.0000 & -0.5196 & -0.1187 & -6.2127 \\ 1.0000 & 0.8470 & 19.0055 & 12.3679 \end{array} \right]$$

or, equivalently,

$$V(s) = \begin{bmatrix} -0.1187 & -0.5196s - 6.2127 \\ s + 19.0055 & 0.8470s + 12.3679 \end{bmatrix},$$

having dimension 2×2 and degree $\partial[V(s)] = 1$.

We must now determine whether the eigenvalues of the solution–assigned closed–loop and feedback compensator polynomial matrices $W_s(s)$ and $V(s)$ are stable. Applying Algorithm 2.3, the characteristic polynomials $w_s(s)$ and $v(s)$ of $W_s(s)$ and $V(s)$ are found to be:

$$w_s(s) = 0.5196s + 8.0618 \text{ and } \quad v(s) = 0.5196s^2 + 15.9881s + 116.6062\,.$$

Hence the positions of the solution–assigned closed–loop and feedback compensator poles are $\{-15.5143\}$ and $\{-18.8859, -11.8818\}$, respectively. These poles, and those of the closed–loop and feedback compensator systems $H(s)$ and $F(s)$, are stable. The set of closed–loop poles are:

$$-3 \pm j4.5, \quad -5 \pm j2, \quad -10 \text{ and } \quad -15.5143$$

and the feedback compensator poles are:

$$-18.8859 \text{ and } \quad -11.8818\,.$$

Thus, a solution to the output–feedback compensator, $F(s)$, is described by

$$\begin{aligned} F(s) &= V^{-1}(s).K(s) \\ &= \begin{bmatrix} -0.1187 & -0.5196s - 6.2127 \\ s + 19.0055 & 0.8470s + 12.3679 \end{bmatrix}^{-1} . \begin{bmatrix} -0.8970 & s + 7.6891 \\ -19.0359 & 25.5093 \end{bmatrix}. \end{aligned}$$

It should be emphasised that a solution to the output–feedback compensator, $F(s)$, is *not* unique. For example, the designer–assigned polynomial matrix,

$$W_d(s) = \begin{bmatrix} 0 & s^3 + 16s^2 + 89.25s + 292.5 \\ s^2 + 10s + 29 & 0 \end{bmatrix},$$

has the same designer–assigned closed–loop pole locations as $W_d(s)$ in eqn. (4.59), yet a different solution for the compensator would have resulted.

Steady–state performance considerations

The denominator $V(s)$ of the LMFD of $R(s) = V^{-1}(s).M(s)$ has already been calculated.

To satisfy the steady-state alignment error specifications, the zero degree numerator $M(s) = M_0$ is determined from the relationship

$$M(0) = M_0 = (J.N(0))^{-1}.w(0).(I_{n_i} - E_{ac}.K_f^{-1}) .$$

Compute the terms $N(0)$, $w(0)$, J, K_f and E_{ac}.

The closed-loop denominator $W(s) = W_s(s).W_d(s)$ is required. Now

$W = W_s. < W_d >_1$

$$= \left[\begin{array}{cc|cc|cc} 0.0000 & -0.5196 & 0.4297 & -10.3928 & 2.5780 & -67.0333 \\ 1.0000 & 0.8470 & 20.8140 & 16.9399 & 118.1339 & 109.2622 \end{array}\right. \cdots$$

$$\cdots \left.\begin{array}{|cc} 12.5679 & -150.6950 \\ 433.3088 & 245.6282 \end{array}\right] .$$

Thus

$$W(0) = W_3 = \begin{bmatrix} 12.5679 & -150.6950 \\ 433.3088 & 245.6282 \end{bmatrix} .$$

Hence

$$w(0) = \det(W(0)) = 68385$$

and

$$\mathrm{Adj}(W(0)) = \begin{bmatrix} 245.6282 & 150.6950 \\ -433.3088 & 12.5679 \end{bmatrix} .$$

Thus

$$\begin{aligned} N(0) &= N_R(0).\mathrm{Adj}(W(0)) \\ &= \begin{bmatrix} 34.4672 & 38.5036 \\ 2.6350 & 3.2628 \\ 27.6569 & 17.9270 \end{bmatrix} . \begin{bmatrix} 245.6282 & 150.6950 \\ -433.3088 & 12.5679 \end{bmatrix} \\ &= \begin{bmatrix} -8217.9 & 5677.9 \\ -766.5 & 438.1 \\ -974.6 & 4393.1 \end{bmatrix} . \end{aligned}$$

From the requirements on the step inputs, the diagonal matrix K_g is set to

$$K_g = \begin{bmatrix} 1 & 0 \\ 0 & 2.5 \end{bmatrix}$$

and

$$J = \begin{bmatrix} 0 & 0 & 1 \\ 1 & 0 & 0 \end{bmatrix} .$$

The column vector of alignment errors e_{ac} is

$$e_{ac} = \begin{bmatrix} 0 \\ 0.8 \end{bmatrix}$$

thus

$$E_{ac} = \begin{bmatrix} 0 & 0 \\ 0 & 0.8 \end{bmatrix}.$$

Hence

$$M_0 = \left(\begin{bmatrix} 0 & 0 & 1 \\ 1 & 0 & 0 \end{bmatrix} . \begin{bmatrix} -8217.9 & 5677.9 \\ -766.5 & 438.1 \\ -974.6 & 4393.1 \end{bmatrix} \right)^{-1} .68385.$$

$$\left(\begin{bmatrix} 1 & 0 \\ 0 & 1 \end{bmatrix} - \begin{bmatrix} 0 & 0 \\ 0 & 0.8 \end{bmatrix} . \begin{bmatrix} 1 & 0 \\ 0 & 2.5 \end{bmatrix}^{-1} \right)$$

$$= \begin{bmatrix} 12.7023 & -6.6830 \\ 18.3845 & -1.4826 \end{bmatrix}.$$

Thus, a solution for the pre-compensator is

$$R(s) = V^{-1}(s).M(s)$$

$$= \begin{bmatrix} -0.1187 & -0.5196s - 6.2127 \\ s + 19.0055 & 0.8470s + 12.3679 \end{bmatrix}^{-1} . \begin{bmatrix} 12.7023 & -6.6830 \\ 18.3845 & -1.4826 \end{bmatrix}.$$

Chapter 5

Design Examples

5.1 Introduction

Four examples of compensator design are presented, not only to illustrate the a design procedure outlined in Chap. 4 but also to demonstrate the application of techniques discussed in the earlier chapters. The examples have been selected from the literature, but the performance specifications may have been altered markedly for illustrative purposes.

1. The first example examines the design of a digital controller for a fourth–order, two–input, two–output state–space model of a chemical reactor which is unstable [101]. The closed–loop system is to be stable, non–oscillatory, non–overshooting and accurate in the steady–state. Decoupling of outputs in the steady state is desirable.

2. The second design example considers a third–order, two–input, three–output transfer function matrix model of a pressurised flow box for a paper machine [43]. The design objective is to develop a continuous–time compensator. The specifications require an improvement in both the speed of the response and the steady–state accuracy of the closed–loop system.

3. The third example deals with the design of a digital controller for a seventh–order, two–input, four–output state–space model of the $L - 1011$ aircraft at cruise flight conditions [2]. For step changes in their reference inputs, the response of the outputs is to be fast, accurate and generally non–interacting.

4. The final example examines the design of a continuous–time output feedback compensator, which implements full output feedback, and a pre–compensator [75]. The plant is a fifth–order, three–input, three–output state–space model of a flight control system. The objective of the design study is to ensure that, for step changes in their reference inputs, the response of the controlled outputs are non–oscillatory, non–overshooting, accurate in the steady–state and generally non–interacting.

5.2 The Chemical Reactor

This example considers the design of a digital output feedback compensator and pre–compensator for a chemical reactor which is unstable [101].
The system is described in state–space form; i.e.,

$$\dot{x}(t) = A.x(t) + B.u(t) \text{ and } c(t) = C.x(t)\,.$$

The plant, which is highly nonlinear, has been analysed to obtain the following linearised state–space model for small perturbations about a selected operating point.

$$A = \begin{bmatrix} 1.380 & -0.2077 & 6.715 & -5.676 \\ -0.5814 & -4.290 & 0 & 0.6750 \\ 1.067 & 4.273 & -6.654 & 5.893 \\ 0.0480 & 4.273 & 1.343 & -2.104 \end{bmatrix}, \quad B = \begin{bmatrix} 0 & 0 \\ 5.679 & 0 \\ 1.136 & -3.146 \\ 1.136 & 0 \end{bmatrix}$$

and

$$C = \begin{bmatrix} 0 & 1 & 0 & 0 \\ 1 & 0 & 1 & -1 \end{bmatrix}.$$

The plant is a two–input, two–output system. The reference inputs $r_1(t)$ and $r_2(t)$ are the material flow rate and the cooling-water flow rate, respectively. The controlled outputs $c_1(t)$ and $c_2(t)$ are the reactor pressure and the sum of the reaction and cooling–jacket temperatures, respectively.

In the s–plane the pole positions of the unstable open–loop reactor are

$$1.991, \quad 0.0632, \quad -5.057 \text{ and } \quad -8.666\,.$$

For illustrative purposes, the closed–loop system specifications are chosen as follows:

- Both the closed–loop system and the compensator are to be stable.
- The four closed–loop poles of the reactor are to be located at -1, -1 and $-3 \pm j$.
- Both the outputs, $c_1(t)$ and $c_2(t)$, are to be used for feedback.
- The steady–state following error for the reactor pressure output, $c_1(t)$, is to be zero when a unit–slope ramp is applied at the material flow rate input, $r_1(t)$.
- For the output $c_2(t)$ the steady–state following error is to be zero when a unit–slope ramp is applied at the cooling–water flow rate reference, $r_2(t)$.
- Steady–state decoupling between the outputs $c_1(t)$ and $c_2(t)$ is required.
- The sampling period, T, is to be 100ms.

The closed–loop control system incorporating the plant is shown in Fig. 5.1.

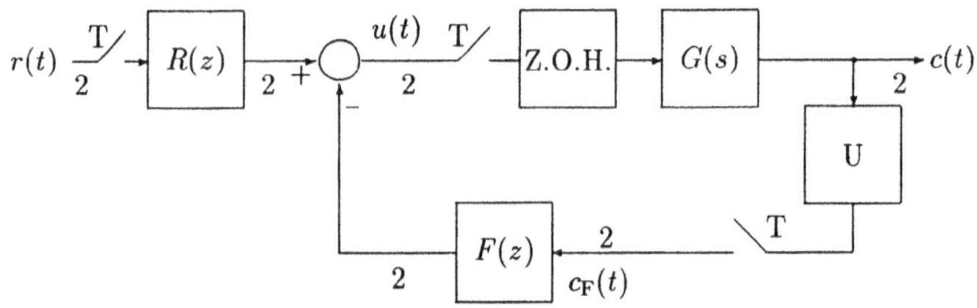

Figure 5.1: Configuration of the closed-loop system.

The SSD $S(A, B, C_F = U.C)$ of the modified plant must be observable for the outputs chosen for feedback. The state space description is therefore reduced to an observable and controllable form. If it is found to be unobservable the choice of outputs selected for feedback must be modified by the designer.

In this example, the SSD of the plant is to be transformed to its discrete–time form, the sampling period being supplied by the designer. The plant matrix A and input matrix

B are transformed to their discrete–time model equivalent A_{D} and B_{D}, respectively, using eqn. (4.47), i.e.

$$A_{\mathrm{D}} = \begin{bmatrix} 1.1782 & 0.0015 & 0.5116 & -0.4033 \\ -0.0515 & 0.6619 & -0.0110 & 0.0613 \\ 0.0762 & 0.3351 & 0.5606 & 0.3824 \\ -0.0006 & 0.3353 & 0.0893 & 0.8494 \end{bmatrix}$$

and

$$B_{\mathrm{D}} = \begin{bmatrix} 0.0045 & -0.0876 \\ 0.4672 & 0.0012 \\ 0.2132 & -0.2353 \\ 0.2131 & -0.0161 \end{bmatrix} .$$

The discrete–time SSD of the plant is now transformed to its right–coprime RMFD by applying Algorithm 3.1. The denominator, D_{R}, and the numerator, N_{R}, are represented by their row block–matrix descriptions,

$$D_{\mathrm{R}} = \left[\begin{array}{cc|cc|cc} 1.0000 & 0.0000 & -1.5790 & -0.0344 & 0.5901 & 0.0143 \\ 0.0000 & 1.0000 & 0.1718 & -1.6712 & -0.0753 & 0.5258 \end{array}\right]$$

having degree $\partial[D_{\mathrm{R}}] = 2$, and

$$N_{\mathrm{R}} = \left[\begin{array}{cc|cc} 0.4672 & 0.0012 & -0.4177 & -0.0112 \\ 0.0046 & -0.3067 & -0.0296 & 0.1853 \end{array}\right] ,$$

of degree $\partial[N_{\mathrm{R}}] = 1$. The interconnection matrix U is

$$U = \begin{bmatrix} 1 & 0 \\ 0 & 1 \end{bmatrix} .$$

The numerator row block–matrix $\bar{N}_{\mathrm{R}} = U.N_{\mathrm{R}}$ is now constructed; in this example $\bar{N}_{\mathrm{R}} = N_{\mathrm{R}}$.

Let us now compute the left minimal basis and extract a coprime, proper and least–order output feedback compensator, $F(z)$.

Before the number of designer–assigned closed–loop poles are selected, the designer is presented with a table which lists the numbers of designer–assigned and solution–assigned closed–loop poles (n_{wd} and n_{ws}, respectively) for a range of feedback compensator orders (n_{v}). The table starts with a zero–order (static) compensator and ends with a compensator order which allows all closed–loop poles to be designer-assigned. At this stage, the numbers of designer– and solution–assigned closed–loop poles are estimates only.

n_v	—	n_{ws}	—	n_{wd}
0		1		3
1		0		5

The problem specification states that four closed–loop poles are to be arbitrarily assigned by the designer. A first–order output feedback compensator permits five closed–loop poles to be designer–assigned. Thus a first–order output–feedback compensator should suffice.

Because the number of designer–assigned closed–loop poles is now five, the additional closed–loop pole is temporarily set at $s = 0$ (or $z = 1$). The effect of a variation in the position of this pole will be examined shortly. The designed–assigned closed–loop pole positions in the s–plane are thus

$$-1, \quad -1, \quad -3 \pm j, \quad 0.$$

These designer–assigned closed–loop pole positions are mapped via $z = e^{sT}$, where T is the sampling period, to their equivalent positions in the z–plane, i.e.

$$0.9048, \quad 0.9048, \quad 0.7371 \pm 0.0740j, \quad 1.$$

Let us plot the loci of both the solution–assigned closed–loop poles and the compensator poles as the fifth designer–assigned closed–loop pole is varied from $z = 1$ to $z = 0.5$ along the real axis. This allows us to gauge the sensitivity of the solution–assigned compensator and closed–loop poles to a variation in the location of a designer–assigned pole.

In Fig. 5.2 are shown the loci of the solution–assigned compensator poles; it is seen that, in fact, there are two solution–assigned compensator loci. Thus a second order, not a first–order, feedback compensator is required to assign the five designer–assigned closed–loop poles exactly. 'X' denotes the beginning and 'O' denotes the end of the loci.

It is apparent that the variation of the fifth designer–assigned closed–loop pole has minimal effect on the loci of the solution–assigned compensator poles. The one locus remains at ~ 0.9, the second locus begins at ~ 0.45 and ends at ~ 0.55.

With a second–order feedback compensator there is now one solution–assigned closed–loop pole. From Fig. 5.2 it is seen that this pole does not shift from ~ 0.58 as the fifth designer–assigned closed–loop pole is varied.

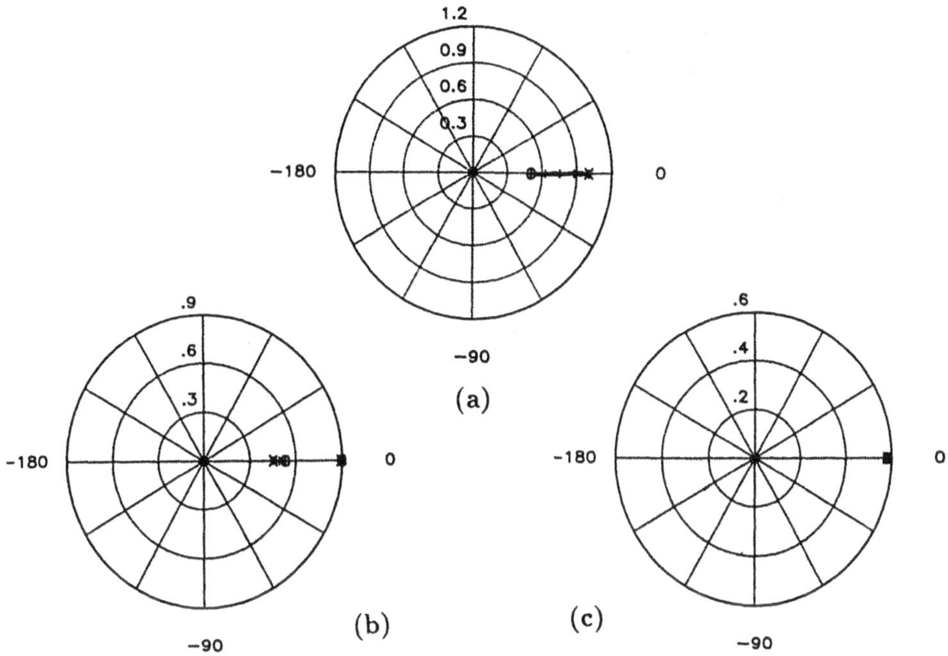

Figure 5.2: Loci in the z–plane of designer–assigned closed–loop pole (a), solution–assigned compensator poles (b), solution–assigned closed–loop pole (c).

In order that the effect on the closed–loop response is minimal, the fifth designer–assigned closed–loop pole is placed at $z = 0.5$. Hence the two solution–assigned compensator poles are located at ~ 0.55 and ~ 0.9, with the solution–assigned closed–loop pole at ~ 0.58. For the five designer–assigned closed–loop pole positions,

$$0.9048, \quad 0.9048, \quad 0.7371 + 0.0740j, \quad 0.7371 - 0.0740j, \quad 0.5\,,$$

the precise solution–assigned compensator pole positions in the z–plane are 0.8962 and 0.5307, the solution–assigned closed–loop pole being located at 0.5790. All solution–assigned poles lie within the unit disc, and thus satisfy the stability requirement. Moreover, their positions within the unit disc are such that their effect on the transient response is likely to be negligible.

In calculating the above solution–assigned poles given the designer–assigned poles, the software implements steps 8 to 14 of Algorithm 4.2 and determines the parameters of the feedback compensator, $F(z)$, without interference from the designer. Some results of these calculations are now given.

For the designer–assigned closed–loop poles, the row block–matrix description W_{d} of the polynomial matrix $W_{\mathrm{d}}(z)$ is computed, i.e.

$$W_{\mathrm{d}} = \left[\begin{array}{cc|cc|cc|cc} 0.0000 & 1.0000 & 0.0000 & -1.9742 & 0.0000 & 1.2859 & 0.0000 & -0.2744 \\ 0.0000 & 0.0000 & 1.0000 & 0.0000 & -1.8096 & 0.0000 & 0.8187 & 0.0000 \end{array}\right],$$

the degree of which is $\partial[W_{\mathrm{d}}] = 3$.

The left minimal basis, $\tilde{Y}(z)$, is given in its block-matrix form, i.e.

$$\tilde{Y} = \left[\begin{array}{cccccc|cccccccccccc} 0 & 0 & 0 & 0 & 0 & 0 & 0 & 0 & 0 & 1 & 0 & -0.9752 & -0.9752 & -2.0832 & 1.2643 & -0.0397 & -1.9203 & 0.6022 \\ 0 & 0 & 0 & 0 & 0 & 0 & 0 & 0 & 1 & 0 & 0 & 0.0865 & 0.0865 & 0.8185 & -1.1136 & -0.1437 & 0.3364 & -0.0275 \\ 0 & 0 & 0 & 0 & 0 & 0 & 0 & 1 & 0 & 0 & 1 & -0.0704 & -0.0704 & -0.7294 & 0.0786 & 0.0240 & -0.9479 & 0.0558 \\ 0 & 0 & 0 & 0 & 0 & 1 & 1 & 0 & 0 & 0 & -0.1718 & -1.6704 & -1.3674 & -0.2386 & 0.1280 & 0.1652 & -0.1476 & 0.6621 \end{array}\right],$$

degree $\partial[\tilde{Y}] = 2$. In this example, $\hat{Y}$ is equal to $\tilde{Y}$. The dimension of the left minimal basis is 4. From $\tilde{Y}$, the row block-matrix $\hat{C}$ is extracted,

$$\hat{C} = \left[\begin{array}{cccc|cccc} 0 & 0 & 0 & 0 & 0 & 1 & 0 & -0.9752 \\ 0 & 0 & 0 & 0 & 1 & 0 & 0 & 0.0865 \\ 0 & 0 & 0 & 0 & 0 & 0 & 1 & -0.0704 \\ 0 & 0 & 0 & 1 & 0 & 0 & -0.1718 & -1.6704 \end{array}\right. \cdots$$

$$\cdots \left.\begin{array}{|cccc} 1.2643 & -0.0397 & -1.9203 & 0.6022 \\ -1.1136 & -0.1437 & 0.3364 & -0.0275 \\ 0.0786 & 0.0240 & -0.9479 & 0.0558 \\ 0.1280 & 0.1652 & -0.1476 & 0.6621 \end{array}\right]$$

having degree $\partial[\hat{C}] = 2$.

From $\hat{C}$ the highest row degree coefficient matrix $\Gamma_r[\hat{C}(z)]$ is formed,

$$\Gamma_r[\hat{C}] = [\widehat{K}_\Gamma \ \hat{V}_\Gamma] = \left[\begin{array}{cc|cc} 0 & 1 & 0 & -0.9752 \\ 1 & 0 & 0 & 0.0865 \\ 0 & 0 & 1 & -0.0704 \\ 0 & 0 & 0 & 1 \end{array}\right].$$

The denominator $V(z)$ and numerator $K(z)$ of the feedback compensator, and the solution-assigned closed-loop polynomial matrix $W_s(z)$ can now be evaluated. The row block-matrices V, K and W_s are extracted from rows 1 and 3 of $\hat{Y}$, yielding:

$$V = \left[\begin{array}{cc|cc} 0.0000 & -0.9752 & -1.9203 & 0.6022 \\ 1.0000 & -0.0704 & -0.9479 & 0.0558 \end{array}\right],$$

degree $\partial[V] = 1$;

$$K = \left[\begin{array}{cc|cc} 0.0000 & 1.0000 & 1.2643 & -0.0397 \\ 0.0000 & 0.0000 & 0.0786 & 0.0240 \end{array}\right],$$

degree $\partial[K] = 1$; and

$$W_s = \left[\begin{array}{cc|cc} 0.0000 & 0.0000 & -0.9752 & -2.0832 \\ 0.0000 & 1.0000 & -0.0704 & -0.7294 \end{array}\right],$$

degree $\partial[W_s] = 1$. The output-feedback compensator is thus given in its LMFD by

$$F(z) = \begin{bmatrix} -1.9203 & -0.9752z + 0.6022 \\ z - 0.9479 & -0.0704z + 0.0558 \end{bmatrix}^{-1} \cdot \begin{bmatrix} 1.2643 & z - 0.0397 \\ 0.0786 & 0.0240 \end{bmatrix}$$

The next phase of the design is to satisfy the steady-state requirements by means of the pre-compensator, $R(z)$.

In the specifications, the steady–state error is a following error. In this case the designer enters the specified following errors and the slopes of the ramp inputs. The steady–state errors of the controlled outputs $c_1(t)$ and $c_2(t)$ are specified with respect to the reference inputs $r_1(t)$ and $r_2(t)$, respectively. The row block–matrix description of the numerator $M(z)$ of the pre–compensator is computed and found to be

$$M = \left[\begin{array}{cc|cc} -4.7995 & 2.3006 & 4.5786 & -1.9828 \\ 0.7224 & 0.0943 & -0.6643 & -0.0775 \end{array}\right],$$

degree $\partial[M] = 1$. The pre–compensator $R(z)$, represented in its LMFD, is thus

$$R(z) =$$

$$\begin{bmatrix} -1.9203 & -0.9752z + 0.6022 \\ z - 0.9479 & -0.0704z + 0.0558 \end{bmatrix}^{-1} \cdot \begin{bmatrix} -4.7995z + 4.5786 & 2.3006z - 1.9828 \\ 0.7224z - 0.6643 & 0.0943z - 0.0775 \end{bmatrix}.$$

Alternatively, the output compensator can be represented by its TFMD,

$$F(z) = \frac{\begin{bmatrix} -0.0123z + 0.0231 & -0.0704z^2 + 0.0820z - 0.0167 \\ -1.2643z + 1.0474 & -z^2 + 0.9876z - 0.0838 \end{bmatrix}}{0.9752z^2 - 1.3915z + 0.4638}$$

and

$$R(z) = \frac{\begin{bmatrix} 1.0423z^2 - 1.6727z + 0.6553 & -0.0699z^2 + 0.1355z - 0.0639 \\ 4.7995z^2 - 10.5152z + 5.6156 & -2.3006z^2 + 3.9825z - 1.7308 \end{bmatrix}}{0.9752z^2 - 1.3915z + 0.4638}.$$

The control algorithm for the digital controller can also be realised. The ARMA model (see § 3.6) for the controller is

$$u(t) = -\overline{V}_1.u(t-1) - \overline{K}_0.c_F(t) - \overline{K}_1.c_F(t-1) + \overline{M}_0.r(t) + \overline{M}_1.r(t-1),$$

where $V = V_0^{-1}.V_i$, $\overline{K}_i = V_0^{-1}.K_i$, and $\overline{M}_i = V_0^{-1}.M_i$. In this example $V(z)$ is regular. (If $V(z)$ is not regular, an algorithm similar to that of Algorithm C.1 is used to form an equivalent control algorithm in which $V(z)$ is regular.)

The control algorithm of the second-order digital compensator design is thus

$$u(t) = -\begin{bmatrix} -0.8093 & 0.0123 \\ 1.9690 & -0.6175 \end{bmatrix}.u(t-1) - \begin{bmatrix} 0.0000 & -0.0722 \\ 0.0000 & -1.0254 \end{bmatrix}.c_F(t)$$

$$-\begin{bmatrix} -0.0126 & 0.0269 \\ -1.2964 & 0.0400 \end{bmatrix}.c_F(t-1) + \begin{bmatrix} 1.0688 & -0.0717 \\ 4.9214 & -2.3590 \end{bmatrix}.r(t)$$

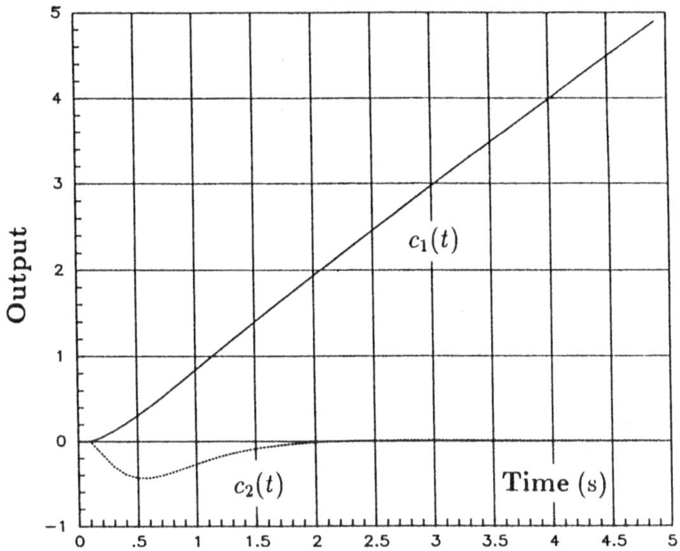

Figure 5.3: Reactor: closed–loop responses to a ramp change in $r_1(t)$.

$$+\begin{bmatrix} -0.9947 & 0.0656 \\ -4.6949 & 2.0332 \end{bmatrix} .r(t-1)\,.$$

The responses of $c_1(t)$ and $c_2(t)$ to a unit slope ramp change in the material flow rate, $r_1(t)$, is shown in Fig. 5.3. The response in the reactor pressure $c_1(t)$ is stable, non–oscillatory, non–overshooting, and accurate; moreover, the interaction is reduced considerably.

The responses of the two outputs for a unit–slope ramp change in the cooling–water flow rate, $r_2(t)$, are shown in Fig. 5.4. These results are acceptable; the response $c_2(t)$ has been achieved with minimal interaction on $c_1(t)$.

5.3 The Pressurised Flow Box

Let us consider the design of a controller for a third–order, two–input, three–output pressurised flow box for a paper machine [43], which is described by the coprime transfer

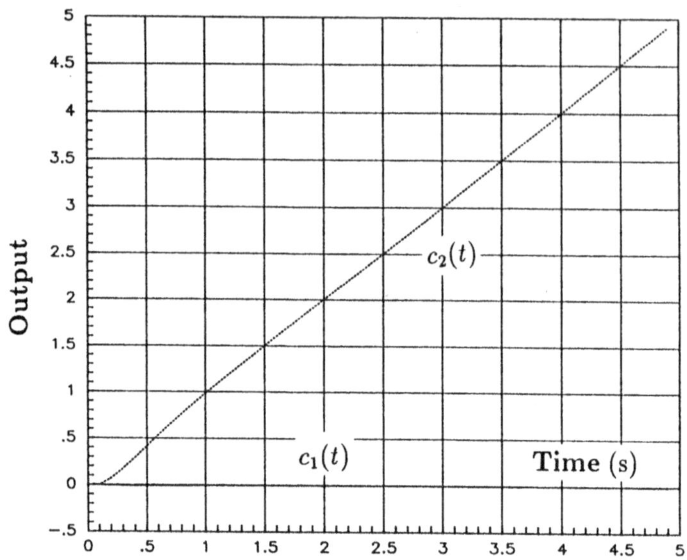

Figure 5.4: Reactor: closed-loop responses to a ramp change in $r_2(t)$.

function matrix

$$G(s) = \frac{\begin{bmatrix} s & s^2 + 1.07s + 0.07 \\ -0.05 & 0.7s^2 + 0.79s + 0.09 \\ s^2 + 0.2s + 0.005 & 0 \end{bmatrix}}{s^3 + 1.2s^2 + 0.205s + 0.005} .$$

The inputs $r_1(t)$ and $r_2(t)$ are the set–points for the air–valve opening and the perturbation in stock valve opening, respectively. The three outputs $c_1(t)$, $c_2(t)$ and $c_3(t)$ are the total head perturbation, stock–level perturbation and the perturbation in air–valve opening, respectively.

A preliminary analysis of the open–loop plant shows that the pole positions in the s–domain are

$$-1, \quad -0.1707 \text{ and } \quad -0.02929 .$$

All the poles lie in the left half of the s–plane, and thus satisfy the stability criterion.

For the purposes of this example, the closed–loop system specifications are chosen as follows:

- The closed–loop system and the output–compensator are to be stable.

- The closed-loop poles are to lie on, or within, the region D of the s-plane, as shown in Fig. 5.5. An initial choice for the closed-loop pole locations is -2 and $-0.4 \pm j0.1$.

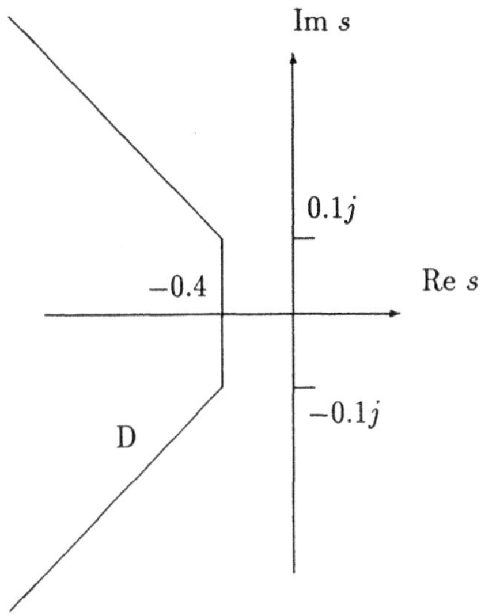

Figure 5.5: Region D for closed-loop pole positions.

- The outputs, $c_1(t)$ and $c_2(t)$, are to be available for feedback.

- The steady-state alignment error for the output $c_1(t)$, with respect to a unit step applied at the input $r_1(t)$, is to be zero.

- The steady-state alignment error for the output $c_2(t)$, with respect to a unit step applied at the input $r_2(t)$, is to be zero.

- Steady-state decoupling between $c_1(t)$ and $c_2(t)$ is required.

The responses of the open-loop plant to a step change applied to the first and second inputs in turn are sluggish and the steady-state accuracy is unacceptable.

The TFMD of the plant $G(s)$ is entered in row block-matrix form. The row block-matrix representation of the numerator $N(s)$ is

$$N = \left[\begin{array}{cc|cc|cc} 0 & 1 & 1 & 1.07 & 0 & 0.07 \\ 0 & 0.7 & 0 & 0.79 & -0.05 & 0.09 \\ 1 & 0 & 0.2 & 0 & 0.005 & 0 \end{array}\right].$$

Similarly, the denominator $d(s)$ is entered in the same form, i.e.

$$d = \left[\begin{array}{cccc} 1 & 1.2 & 0.205 & 0.005 \end{array} \right].$$

The TFMD is first transformed to its SSD, which is checked for controllability and observability. If the SSD is found to be uncontrollable and/or unobservable, it is reduced to one which is both controllable and observable using routines available in commercial software packages.

At this stage the SSD of the plant is transformed by the software to its right–coprime RMFD. The denominator D_{R} and the numerator N_{R}, represented in their row block–matrix forms, are

$$D_{\mathrm{R}} = \left[\begin{array}{cc|cc|cc} 1.0000 & 0.0000 & 1.1286 & 0.0586 & 0.1286 & 0.0586 \\ 0.0000 & 0.0000 & 0.0000 & 1.0000 & 0.0714 & 0.0714 \end{array} \right],$$

degree $\partial[D_{\mathrm{R}}] = 2$, and

$$N_{\mathrm{R}} = \left[\begin{array}{cc|cc} 0.0000 & 0.0000 & 1.0000 & 1.0000 \\ 0.0000 & 0.0000 & 0.0000 & 0.7000 \\ 1.0000 & 0.0000 & 0.1286 & 0.0586 \end{array} \right],$$

degree $\partial[N_{\mathrm{R}}] = 1$. The feedback interconnection matrix is

$$U = \left[\begin{array}{ccc} 1 & 0 & 0 \\ 0 & 1 & 0 \end{array} \right].$$

The table of designer–assigned/solution–assigned closed–loop configurations for a range of compensator orders is

n_v	—	n_{ws}	—	n_{wd}
0		0		3

In the problem specification it states that the three plant poles are to be arbitrarily positioned by the designer. A zero–order (static) output feedback compensator satisfies this requirement.

As specified, the designer–assigned closed–loop poles are to be -2, $-0.4 \pm 0.1j$.

In proceeding to solve for the feedback compensator it is found that a first–order feedback compensator is required to assign the three designer–assigned closed–loop poles exactly, and thus one solution–assigned compensator pole and one solution–assigned closed–loop pole is introduced. Moreover, the solution–assigned closed–loop pole which lies at $s = 0.872$ is unstable.

To assess the effect on the solution–assigned closed–loop pole of a shift in a designer–assigned closed–loop pole location, the pole–locus plotting facility is employed. The designer–assigned, real closed–loop pole is used as variable parameter.

As shown in Fig. 5.6, the real pole located at $s = -2$ is shifted along the real axis to the origin ('X' denotes the beginning and 'O' the end of the loci). The locus of the solution–assigned compensator shows that the compensator pole, located at ~ -0.728, is unaffected by the shift of the designer–assigned pole. However, from the same figure it is apparent that it has a significant effect on the solution–assigned closed–loop pole. As the designer–assigned pole is shifted towards the origin the solution–assigned closed–loop pole moves towards the left–half of the s–plane. It is therefore necessary to relocate the designer–assigned, real closed–loop pole from -2 to -0.6, say. This pole relocation causes the solution–assigned closed–loop pole to shift to ~ -0.53, which lies within the specified region of the s–plane. Furthermore, when the designer–assigned pole is located at -0.6, the solution–assigned compensator pole is stable, being positioned at ~ -0.73. For designer–assigned closed–loop poles, -0.6, $-0.4 \pm 0.1j$, the precise locations of the solution–assigned compensator pole and closed–loop pole are -0.7286 and -0.5286, respectively.

The software then proceeds to calculate the feedback–compensator parameters. Some intermediate results are given below.

The row block–matrix $W_{\rm d}$ of the designer–assigned, closed–loop polynomial matrix, $W_{\rm d}(s)$, is

$$W_{\rm d} = \left[\begin{array}{cc|cc|cc} 0.0000 & 1.0000 & 0.0000 & 0.80000 & 0.0000 & 0.1700 \\ 0.0000 & 0.0000 & 1.0000 & 0.0000 & 0.6000 & 0.0000 \end{array} \right],$$

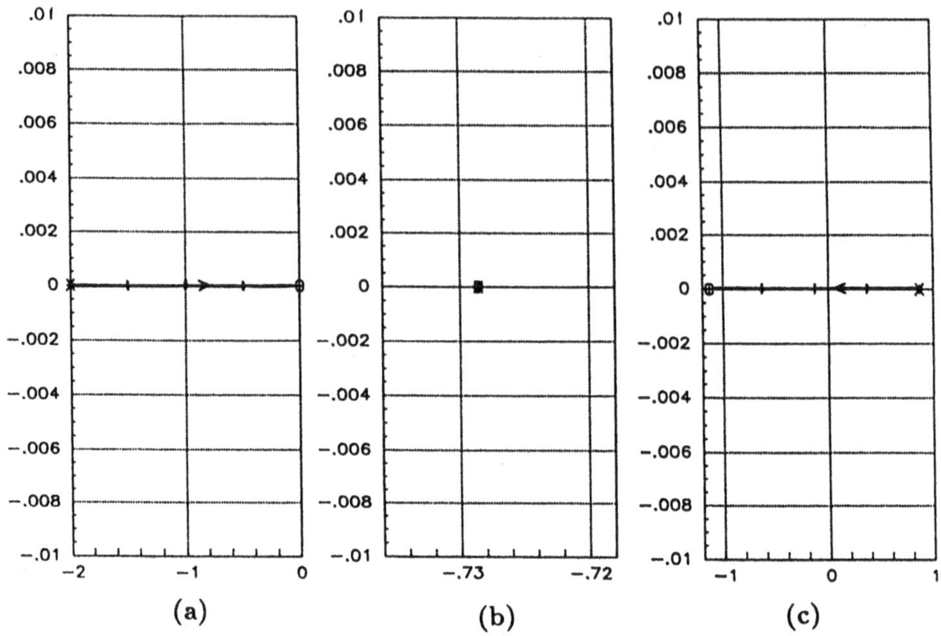

Figure 5.6: Loci in the s–plane of designer–assigned closed–loop pole (a), solution assigned compensator pole (b), solution–assigned closed–loop pole (c).

of degree $\partial[W_d] = 2$. The row–block matrices $\tilde{Y}$, $\hat{Y}$ are

$$\tilde{Y} = \hat{Y} = \left[\begin{array}{cccccc|} 0 & 0 & 0 & 0 & 0 & 1 \\ 0 & 0 & 0 & 1 & 0 & 0 \\ 0 & 0 & 1 & 0 & 0 & 0 \\ 0 & 1 & 0 & 0 & 0 & 0 \end{array}\right. \cdots$$

$$\cdots \left.\begin{array}{cccccc} 1 & 0.0714 & -0.0092 & 0.1816 & 0 & 0.7286 \\ 0 & 0 & 0.0500 & 0 & 0 & -0.7000 \\ 0 & 1 & 0.6714 & -0.8571 & 0 & -1 \\ 0 & 0.5286 & 0.1928 & -0.3531 & 1 & -0.0586 \end{array}\right],$$

degree $\partial[\tilde{Y}] = 1$. The dimension of the left minimal basis is 4.

$$\hat{C} = \left[\begin{array}{cccc|cccc} 0 & 0 & 0 & 1 & -0.0092 & 0.1816 & 0 & 0.7286 \\ 0 & 1 & 0 & 0 & 0.0500 & 0 & 0 & -0.7000 \\ 1 & 0 & 0 & 0 & 0.6714 & -0.8571 & 0 & -1 \\ 0 & 0 & 0 & 0 & 0.1928 & -0.3531 & 1 & -0.0586 \end{array}\right],$$

of degree $\partial[\hat{C}] = 1$, and

$$\Gamma_r[\hat{C}] = \left[\begin{array}{cc|cc} 0 & 0 & 0 & 1 \\ 0 & 1 & 0 & 0 \\ 1 & 0 & 0 & 0 \\ 0.1928 & -0.3531 & 1 & -0.0586 \end{array}\right].$$

The row block–matrices V, K and W_s are extracted from rows 1 and 4 of $\hat{Y}$, yielding

$$V = \left[\begin{array}{cc|cc} 0.0000 & 1.0000 & 0.0000 & 0.7286 \\ 0.0000 & 0.0000 & 1.0000 & -0.0586 \end{array}\right],$$

degree $\partial[V] = 1$;

$$K = \left[\begin{array}{cc} -0.0092 & 0.1816 \\ 0.1928 & -0.3531 \end{array}\right],$$

degree $\partial[K] = 0$; and

$$W_s = \left[\begin{array}{cc|cc} 0.0000 & 0.0000 & 1.0000 & 0.0714 \\ 0.0000 & 1.0000 & 0.0000 & 0.5286 \end{array}\right],$$

degree $\partial[W_s] = 1$.

The pre–compensator $R(s)$ is now designed to satisfy the steady–state requirements. The alignments errors of the controlled outputs $c_1(t)$ and $c_2(t)$ are specified with respect to a unit step change at the reference inputs, $r_1(t)$ and $r_2(t)$, respectively. The row block–matrix, M, of the numerator, $M(s)$, of the pre–compensator, $R(s)$, is calculated to be

$$M = \begin{bmatrix} 0.0429 & 0.1816 \\ 0.3171 & -0.4531 \end{bmatrix},$$

having degree $\partial[M] = 0$.

The output compensator which results from the design procedure is

$$F(s) = \begin{bmatrix} 0 & s+0.7286 \\ 1 & -0.0586 \end{bmatrix}^{-1} \cdot \begin{bmatrix} -0.0092 & 0.1816 \\ 0.1928 & -0.3531 \end{bmatrix}$$

and

$$R(s) = \begin{bmatrix} 0 & s+0.7286 \\ 1 & -0.0586 \end{bmatrix}^{-1} \cdot \begin{bmatrix} 0.0429 & 0.1816 \\ 0.3171 & -0.4531 \end{bmatrix}.$$

The responses, $c_1(t)$ and $c_2(t)$, for a unit step change in air–valve opening, $r_1(t)$, are shown in Fig. 5.7. The specified steady–state static accuracy and the improved speed of response of the output $c_1(t)$ is achieved with a small transient variation in $c_2(t)$.

The responses of $c_1(t)$ and $c_2(t)$ to a unit step change in stock–valve opening, $r_2(t)$, are shown in Fig. 5.8. These results again appear to be acceptable, the response of $c_2(t)$ being faster and the interaction being considerably reduced compared with that in the open–loop condition.

5.4 The L-1011 Aircraft

The third example considers the design of a digital controller for a seventh–order, two–input, four–output lateral–axis model of the L-1011 aircraft at cruise flight conditions [2]. The model includes actuator dynamics and a washout filter on the yaw rate.

The plant, the aircraft at cruise flight conditions, is described by a set of linearised equations of motion for the lateral dynamics. The state variables, the reference inputs and output variables are given by eqns. (5.1), (5.2) and (5.3), respectively.

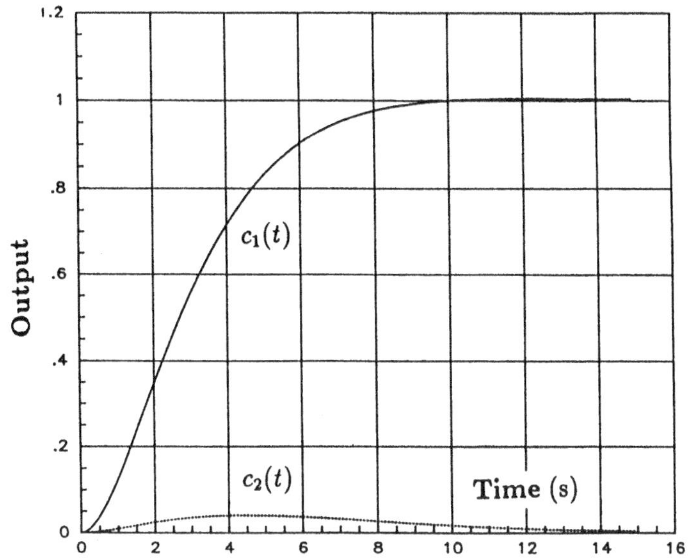

Figure 5.7: Flow box: closed–loop responses to a step change in $r_1(t)$.

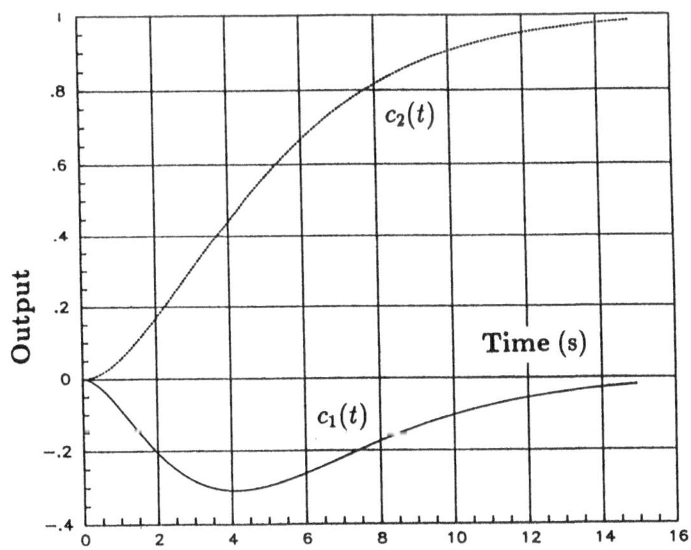

Figure 5.8: Flow box: closed–loop responses to a step change in $r_2(t)$.

The state vector is

$$x(t) = \begin{bmatrix} \delta_{\mathrm{r}}(t) \\ \delta_{\mathrm{a}}(t) \\ \varphi(t) \\ y(t) \\ \rho(t) \\ \beta(t) \\ x_7(t) \end{bmatrix} \quad \begin{matrix} \text{rudder deflection (rad)} \\ \text{aileron deflection (rad)} \\ \text{bank angle (rad)} \\ \text{yaw rate (rad s}^{-1}\text{)} \\ \text{roll rate (rad s}^{-1}\text{)} \\ \text{sideslip angle (rad)} \\ \text{washout filter state} \end{matrix} \quad , \tag{5.1}$$

the reference input vector is

$$r(t) = \begin{bmatrix} \delta_{\mathrm{rc}}(t) \\ \delta_{\mathrm{ac}}(t) \end{bmatrix} \quad \begin{matrix} \text{rudder command (rad)} \\ \text{aileron command (rad)} \end{matrix} \quad , \tag{5.2}$$

and the output vector is given by

$$c(t) = \begin{bmatrix} r_{\mathrm{wo}}(t) \\ \rho(t) \\ \beta(t) \\ \varphi(t) \end{bmatrix} \quad \begin{matrix} \text{yaw rate with washout} \\ \text{roll rate} \\ \text{sideslip angle} \\ \text{bank angle} \end{matrix} \tag{5.3}$$

The aircraft at cruise flight conditions is described by the linearised SSD,

$$A = \begin{bmatrix} -20 & 0 & 0 & 0 & 0 & 0 & 0 \\ 0 & -25 & 0 & 0 & 0 & 0 & 0 \\ 0 & 0 & 0 & 0 & 1 & 0 & 0 \\ -0.744 & -0.032 & 0 & -0.154 & -0.0042 & 1.54 & 0 \\ 0.337 & -1.12 & 0 & 0.249 & -1.0 & -5.2 & 0 \\ 0.02 & 0 & 0.0386 & -0.996 & -0.000295 & -0.117 & 0 \\ 0 & 0 & 0 & 0.5 & 0 & 0 & -0.5 \end{bmatrix},$$

$$B = \begin{bmatrix} 20 & 0 \\ 0 & 25 \\ 0 & 0 \\ 0 & 0 \\ 0 & 0 \\ 0 & 0 \\ 0 & 0 \end{bmatrix} \quad \text{and } C = \begin{bmatrix} 0 & 0 & 0 & 1 & 0 & 0 & -1 \\ 0 & 0 & 0 & 0 & 1 & 0 & 0 \\ 0 & 0 & 0 & 0 & 0 & 1 & 0 \\ 0 & 0 & 1 & 0 & 0 & 0 & 0 \end{bmatrix}.$$

The plant poles are located in the s–plane at

$$-0.0884 \pm j1.272, \quad -1.085, \quad -0.0091, \quad -0.5, \quad -20 \text{ and } \quad -25\,.$$

A digital output feedback compensator and pre–compensator are to be designed such that the following specifications are satisfied:

- The closed-loop system and the compensator are to be stable.
- The filtered yaw rate, $r_{wo}(t)$, roll rate, $\rho(t)$, and bank angle, $\varphi(t)$, are to be available for feedback.
- For closed-loop operation it is desirable that all the poles lie within a disc of radius 0.94 in the z-plane.
- The steady-state alignment error for the bank angle, $\varphi(t)$, with respect to a one degree step applied at the aileron command, $\delta_{ac}(t)$, is to be zero.
- The steady-state alignment error for the sideslip angle, $\beta(t)$, with respect to a one degree step applied at the rudder command, $\delta_{rc}(t)$, is to be zero.
- Steady-state decoupling between the bank angle, $\varphi(t)$, and the sideslip angle, $\beta(t)$, is required.
- The sampling period, T, is to be 50 ms.

The design specifications listed above are used only for illustrative purposes and do not purport to represent any actual design specifications for this or any other aircraft.

The pole positions of the open-loop plant in the z-plane are

$$0.9976 \pm 0.0635j, \quad 0.9472, \quad 0.9995, \quad 0.9753, \quad 0.3679 \text{ and } \quad 0.2865\,.$$

It is required that the plant poles $0.9976 \pm 0.0635j$, 0.9472, 0.9995 and 0.9753 be shifted to lie within the disc of radius 0.94.

The pole shifting technique proposed is as follows:

1. Multiply the five poles to be shifted by a factor α, where $0 < \alpha < 1$.
2. Examine the locations of the solution-assigned closed-loop poles and/or compensator poles as α is varied over a range of values. Determine the value of α such that all poles lie within the specified disc.

The software checks that the matrices of the SSD are dimensionally correct and that the model is controllable and observable. The system is to be transformed to the discrete–time domain; the sampling period is $T = 50$ ms. The plant is described by the discrete–time SSD, the plant matrix A_{D} and input matrix B_{D} being

$$A_{\mathrm{D}} = \begin{bmatrix} 0.3679 & 0 & 0 & 0 & 0 & 0 & 0 \\ 0 & 0.2865 & 0 & 0 & 0 & 0 & 0 \\ 0.0003 & -0.0009 & 1 & 0.0004 & 0.0488 & -0.0064 & 0 \\ -0.0234 & -0.0009 & 0.0001 & 0.9904 & -0.0002 & 0.0765 & 0 \\ 0.0100 & -0.0310 & -0.0002 & 0.0184 & 0.9512 & -0.2522 & 0 \\ 0.0013 & 0 & 0.0019 & -0.0494 & 0 & 0.9923 & 0 \\ -0.0003 & 0 & 0 & 0.0246 & 0 & 0.0010 & 0.9753 \end{bmatrix}$$

and

$$B_{\mathrm{D}} = \begin{bmatrix} 0.6321 & 0 \\ 0 & 0.7135 \\ 0.0001 & -0.0004 \\ -0.0136 & -0.0007 \\ 0.0060 & -0.0236 \\ 0.0006 & 0 \\ -0.0001 & 0 \end{bmatrix}.$$

The RMFD of the discrete–time plant is

$$D_{\mathrm{R}} = \left[\begin{array}{cc|cc|cc} 1.0000 & 0.0000 & -3.3230 & 0.1000 & 4.0020 & -0.2331 \\ 0.0000 & 0.0000 & 0.0000 & 1.0000 & -0.0059 & -2.2406 \end{array} \cdots \right.$$

$$\left. \cdots \begin{array}{|cc|cc} -2.0320 & 0.1692 & 0.3530 & -0.0357 \\ 0.0075 & 1.5130 & -0.0016 & -0.2731 \end{array} \right],$$

having degree $\partial[D_{\mathrm{R}}] = 4$, and

$$N_{\mathrm{R}} = \left[\begin{array}{cc|cc|cc} -0.0135 & 0.0000 & 0.0173 & -0.0020 & 0.0057 & 0.0006 \\ 0.0060 & 0.0000 & -0.0082 & -0.0230 & -0.0022 & 0.0081 \\ 0.0006 & 0.0000 & 0.0001 & 0.0001 & -0.0008 & 0.0001 \\ 0.0001 & 0.0000 & 0.0002 & -0.0004 & -0.0003 & -0.0013 \end{array} \cdots \right.$$

$$\left. \cdots \begin{array}{|cc} -0.0095 & 0.0014 \\ 0.0045 & 0.0150 \\ 0.0001 & 0.0000 \\ -0.0001 & -0.0002 \end{array} \right],$$

of degree $\partial[N_{\mathrm{R}}] = 3$. In the problem specification it is stated that the filtered yaw rate, $r_{\mathrm{wo}}(t)$, roll rate, $\rho(t)$, and bank angle, $\varphi(t)$, are to be available for feedback. The interconnection matrix is thus

$$U = \begin{bmatrix} 1 & 0 & 0 & 0 \\ 0 & 1 & 0 & 0 \\ 0 & 0 & 0 & 1 \end{bmatrix} .$$

The table below lists the numbers of designer–assigned and solution–assigned closed–loop poles ($n_{\rm wd}$ and $n_{\rm ws}$, respectively) for a range of feedback compensator orders ($n_{\rm v}$). The table starts with a zero–order (static) compensator and ends with a third–order compensator which allows all closed–loop poles to be designer–assigned. At this stage, the numbers of designer– and solution–assigned closed–loop poles are estimates only.

$n_{\rm v}$	— $n_{\rm ws}$ —	$n_{\rm wd}$
0	3	4
1	2	6
2	1	8
3	0	10

In the problem specification it is required that five closed–loop poles be arbitrarily assigned by the designer. A first–order output feedback compensator from the above table allows for six closed–loop poles to be designer–assigned and is thus selected.

Table 5.1 lists the solution–assigned closed–loop poles and compensator poles for values of α between 0.94 and 0.84. It is found that a second–order compensator is required to assign the five designer–assigned closed–loop poles exactly. A pattern in the movement of the solution–assigned poles can be observed from the table.

Firstly, we notice that as α is reduced the unstable compensator pole for $\alpha = 0.94$ shifts towards the origin of the z–plane. The solution–assigned closed–loop poles and the compensator poles all lie within the unit disc for $0.84 \leq \alpha \leq 0.93$, and thus satisfy the stability criterion. However, for a given value of α, all but one solution–assigned closed–loop pole lie in the specified disc of radius 0.94. Futhermore, this pole moves away from the origin of the z–plane as α is reduced. Thus a compromise between satisfying the specifications and what can actually be achieved is required.

From Table 5.1 it is seen that the most acceptable positions for the solution–assigned poles, such that the closed–loop specifications are realised as closely as possible, is for

α	Closed–loop pole locations	Compensator pole locations
0.94	0.9454, 0.5671 0.2400, −0.0013	1.0010, −0.054
0.93	0.9528, 0.5823 0.2315, −0.0012	0.9793, −0.0575
0.92	0.9571, 0.6019 0.2215, −0.0012	0.9592, −0.0618
0.90	0.9619, 0.6444 0.1953, −0.0016	0.9199, −0.0686
0.88	0.9644, 0.6856 0.1577, −0.0023	0.8797, −0.0709
0.86	0.9659, 0.7229 0.1033, −0.0043	0.8377, −0.0682
0.84	0.9668, 0.7557 0.0313, −0.0175	0.7943, −0.0607

Table 5.1: The positions of the solution–assigned poles for $\alpha = 0.94$ to $\alpha = 0.84$.

$\alpha = 0.93$. The designer–assigned closed–loop poles for this value of α are located in the z–plane at

$$0.9278 \pm 0.0591j, \quad 0.8809, \quad 0.9295 \text{ and } \quad 0.9070\,.$$

The row block–matrix description, $W_{\rm d}$, of the designer–assigned closed–loop polynomial matrix, $W_{\rm d}(z)$, is

$$W_{\rm d} = \left[\begin{array}{cc|cc|cc|cc} 0 & 1 & 0 & -2.7626 & 0 & 2.5473 & 0 & -0.7839 \\ 0 & 0 & 1 & 0 & -1.8104 & 0 & 0.8188 & 0 \end{array}\right],$$

having degree $\partial[W_{\rm d}] = 3$.

The row block–matrices $\tilde{Y}$, $\hat{Y}$, $\widehat{C}$, $\Gamma_{\rm r}[\widehat{C}]$, V, K and $W_{\rm s}$ are given below. For this example $\tilde{Y}$ and $\hat{Y}$ are not equal, that is, the submatrix $[\widehat{K}(z)\ \hat{V}(z)]$ is not row–proper. The unimodular row block–matrix $U_{\rm P}$ of degree 2 premultiplies $\tilde{Y}$ to realize $\hat{Y}$ and $\widehat{C}$; the submatrix $[\widehat{K}(z)\ \hat{V}(z)]$ is then row–proper.

$$\tilde{Y} = \left[\begin{array}{ccccccc|ccccccc} 0 & 0 & 0 & 0 & 0 & 0 & 0 & 0 & 0 & 1 & -14 & 530 & 0 & 0 \\ 0 & 0 & 0 & 0 & 0 & 0 & 0 & 0 & 1 & 0 & -1168 & 43727 & 0 & 0 \\ 0 & 0 & 0 & 0 & 0 & 0 & 0 & 1 & 0 & 0 & -30 & 1032 & 0 & 1 \\ 0 & 0 & 0 & 0 & 1 & 0 & 0 & 0 & 0 & 0 & 0 & 3 & 0 & 0 \\ 0 & 0 & 0 & 1 & 0 & 0 & 0 & 0 & 0 & 0 & -6 & 159 & 0 & 0 \end{array}\right. \cdots$$

$$
\cdots \left| \begin{array}{ccccccc}
0 & 0 & -1 & -13 & -531 & 0 & 0 \\
7 & -1 & -6 & -1023 & -43728 & 2 & -2 \\
0 & 0 & 0 & -22 & -1032 & 0 & -1 \\
0 & 0 & 0 & 0 & -4 & 0 & 0 \\
0 & 0 & 0 & -3 & -159 & 0 & 0
\end{array} \right],
$$

degree $\partial[Y] = 2$. The dimension of the left minimal basis is 5.

$$
U_{\mathrm{P}} = \left[\begin{array}{ccccc|cccc}
0 & 0 & 0 & 0 & 0 & 0 & 0 & 0 & 0 \\
0 & 0 & 0 & 0 & 0 & 0 & 0 & 0 & 0 \\
0 & 0 & 0 & 0 & 0 & 0 & 0 & 0 & 0 \\
0 & 0 & 0 & 0 & 0 & 0 & -2.2869 \times 10^{-5} & 0 & 0 \\
0 & 0.09655 & 0 & 0 & 0 & 0.6264 & -0.0571 & -0.6409 & -4221.8
\end{array} \right. \cdots
$$

$$
\cdots \left. \begin{array}{c|ccccc}
0 & 1 & 0 & 0 & 0 & 0 \\
0 & 0 & 1 & 0 & 0 & 0 \\
0 & 0 & 0 & 1 & 0 & 0 \\
0 & 0 & 0 & 0 & 1 & -0.0267 \\
112.738 & 0 & 0 & 0 & 0 & 1
\end{array} \right],
$$

degree $\partial[U_{\mathrm{P}}] = 2$.

$$
\widehat{Y} = \left[\begin{array}{ccccccc|ccccccc}
0 & 0 & 0 & 0 & 0 & 0 & 0 & 0 & 0 & 0 & 0 & 0 & 0 & 0 \\
0 & 0 & 0 & 0 & 0 & 0 & 0 & 0 & -2.287 \times 10^{-5} & 0 & 0 & 0 & 0 & 0 \\
0 & 0.09655 & 0 & 0 & 0 & 0 & 0 & 0 & -0.1479 & 0 & 0 & 0 & 0 & 0 \\
0 & 0 & 0 & 0 & 0 & 0 & 0 & 0 & 0 & 0 & 0 & 0 & 0 & 0 \\
0 & 0 & 0 & 0 & 0 & 0 & 0 & 0 & 0 & 0 & 0 & 0 & 0 & 0
\end{array} \right. \cdots
$$

$$
\cdots \left| \begin{array}{cccccc}
1 & 0 & 0 & -30.495 & 1031.78 & 0 \\
-1.5181 \times 10^{-4} & 2.151 \times 10^{-5} & 1.4837 \times 10^{-4} & 0.01863 & -0.6688 & 0 \\
-0.03887 & 0.05309 & -0.2849 & -4.9968 & 159.08 & 0.0965 \\
0 & 0 & 1 & -13.595 & 530.47 & 0 \\
0 & 1 & 0 & -1168 & 43727 & 0
\end{array} \right. \cdots
$$

$$
\cdots \begin{array}{c|ccc}
1 & -0.2339 & 2.4798 \times 10^{-3} & -0.2344 \\
-1.5181 \times 10^{-4} & -2.9028 \times 10^{-5} & -2.0116 \times 10^{-7} & 4.5695 \times 10^{-5} \\
-0.0485 & 3.6966 \times 10^{-3} & 2.3886 \times 10^{-5} & -5.212 \times 10^{-3} \\
0 & 0.1076 & 2.6945 \times 10^{-4} & -0.9776 \\
0 & 6.6383 & -0.9406 & -6.4878
\end{array} \cdots
$$

$$
\cdots \left. \begin{array}{cccc}
-21.561 & -1032.1 & 0.0704 & -1.029 \\
0.0149 & 0.6688 & -5.9688 \times 10^{-5} & 2.0282 \times 10^{-4} \\
-2.964 & -159 & 6.9388 \times 10^{-3} & -0.0242 \\
-12.97 & -531 & 0.0373 & 0.0170 \\
-1023.3 & -43728 & 2.2379 & -1.9971
\end{array} \right],
$$

degree $\partial[\hat{Y}] = 3$.

$$\hat{C} = \left[\begin{array}{ccccc|c} 0 & -30.4951 & 1031.78 & 0 & 1 & \\ 1.4837\times10^{-4} & 0.01863 & -0.6688 & 0 & -1.5181\times10^{-4} & \\ -0.2849 & -4.9968 & 159.08 & 0.0965 & -0.04853 & \cdots \\ 1 & -13.595 & 530.47 & 0 & 0 & \\ 0 & -1168 & 43727 & 0 & 0 & \end{array}\right.$$

$$\left.\begin{array}{cccccc} & -0.2344 & -21.561 & -1032.1 & 0.0704 & -1.029 \\ & 4.5695\times10^{-5} & 0.01494 & 0.6688 & -5.9688\times10^{-5} & 2.0282\times10^{-4} \\ \cdots & -5.212\times10^{-3} & -2.964 & -159.01 & 6.9388\times10^{-3} & -0.0242 \\ & -0.9776 & -12.973 & -530.52 & 0.0373 & 0.0170 \\ & -6.4878 & -1023.34 & -43728 & 2.2379 & -1.9971 \end{array}\right],$$

degree $\partial[\hat{C}] = 1$. The highest row degree coefficient matrix $\Gamma_r[\hat{C}(z)]$ is, in row block-matrix form,

$$\Gamma_r[\hat{C}] = \left[\begin{array}{ccc|cc} 0 & -30.4951 & 1031.78 & 0 & 1 \\ 1.4837\times10^{-4} & 0.01863 & -0.6688 & 0 & -1.5181\times10^{-4} \\ -0.2849 & -4.9964 & 159 & 0.0965 & -0.04853 \\ 1 & -13.595 & 530.47 & 0 & 0 \\ 0 & -1168 & 43727 & 0 & 0 \end{array}\right].$$

The row block-matrices V, K and W_s are extracted from rows 1 and 3 of $\hat{Y}$, yielding:

$$V = \left[\begin{array}{cc|cc} 0 & 1 & 0.0704 & -1.0291 \\ 0.0965 & -0.04853 & 0.00694 & -0.0242 \end{array}\right],$$

degree $\partial[V] = 1$;

$$K = \left[\begin{array}{ccc|ccc} 0 & -30.4951 & 1031.78 & -0.2344 & -21.561 & -1032.1 \\ -0.2849 & -4.9964 & 159.08 & -0.00521 & -2.964 & -159.01 \end{array}\right],$$

degree $\partial[K] = 1$; and

$$W_s = \left[\begin{array}{cc|cc|cc} 0 & 0 & 0 & 0 & 1 & 0 \\ 0 & 0.0965 & 0 & -0.1479 & -0.0389 & 0.0531 \end{array}\right. \cdots$$

$$\cdots \left|\begin{array}{cc} -0.2339 & 0.0025 \\ 0.0037 & 2.3886\times10^{-5} \end{array}\right],$$

degree $\partial[W_s] = 3$.

An steady-state error of zero degrees is required for both the sideslip angle and bank angle for a 1 degree step change at the rudder and aileron reference inputs. The row block-matrix, M, of the numerator of the pre-compensator $R(z)$ is calculated and found to be

$$M = \begin{bmatrix} 0.2746 & -0.3018 \\ 0.5231 & 0.0626 \end{bmatrix},$$

degree $\partial[M] = 0$.

The LMFDs of the output-feedback compensator and the pre-compensator are

$$F(z) = \begin{bmatrix} 0.0704 & z - 1.0291 \\ 0.0965z + 0.00694 & -0.04853z - 0.0242 \end{bmatrix}^{-1} .$$

$$\begin{bmatrix} -0.2344 & -3.04915z - 21.561 & 1031.8z - 1032.1 \\ -0.2849z + 0.00521 & -4.9964z - 2.964 & 159.08z - 159.01 \end{bmatrix}$$

and

$$R(z) = \begin{bmatrix} 0.0704 & z - 1.0291 \\ 0.0965z + 0.00694 & -0.04853z - 0.0242 \end{bmatrix}^{-1} . \begin{bmatrix} 0.2746 & -0.3018 \\ 0.5231 & 0.0626 \end{bmatrix} .$$

The control algorithm for the digital compensator can also be realised. The ARMA model for the controller is

$$u(t) = -\overline{V}_1.u(t-1) - \overline{K}_0.c_F(t) - \overline{K}_1.c_F(t-1) + \overline{M}_0.r(t-1),$$

where $\overline{V} = V_0^{-1}.V_i$, $\overline{K}_i = V_0^{-1}.K_i$, and $\overline{M}_i = V_0^{-1}.M_i$. In this example $V(z)$ is regular. The control algorithm for the second-order digital compensator design is

$$u(t) = -\begin{bmatrix} 0.1073 & -0.7682 \\ 0.0704 & -1.0291 \end{bmatrix} .u(t-1) - \begin{bmatrix} -2.9507 & -67.0782 & 2166.2537 \\ 0 & -30.4952 & 1031.7845 \end{bmatrix} .c_F(t)$$

$$- \begin{bmatrix} -0.1718 & -41.5398 & -2165.743 \\ -0.2344 & -21.5613 & -1032.0854 \end{bmatrix} .c_F(t-1) + \begin{bmatrix} 5.5556 & 0.4970 \\ 0.2746 & -0.3018 \end{bmatrix} r(t-1) .$$

In closed loop, the responses of the sideslip angle ($c_3(t) = \beta(t)$) and bank angle ($c_4(t) = \varphi(t)$) when 1 degree (or 0.01745 rad) step changes are applied to the rudder command and aileron command inputs in turn are shown in Fig. 5.9 and 5.10, respectively. The speed of response of the closed-loop system is improved and the desired steady-state static accuracy and decoupling are achieved.

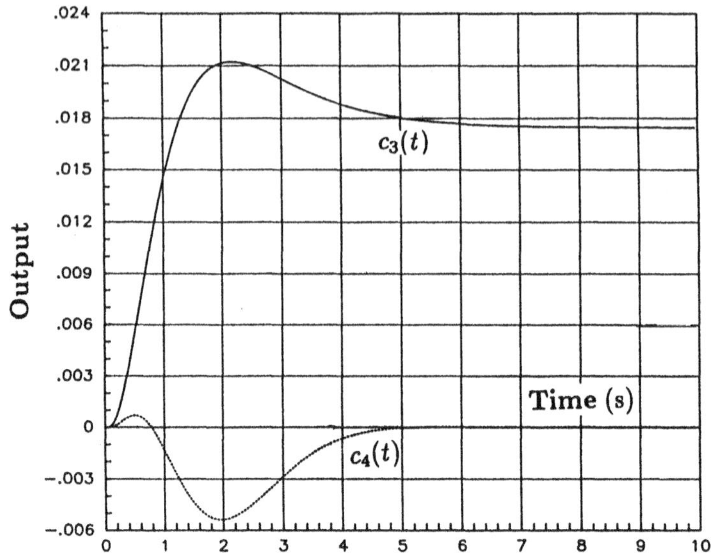

Figure 5.9: L-1011 aircraft: closed–loop responses to a step change in $r_1(t)$.

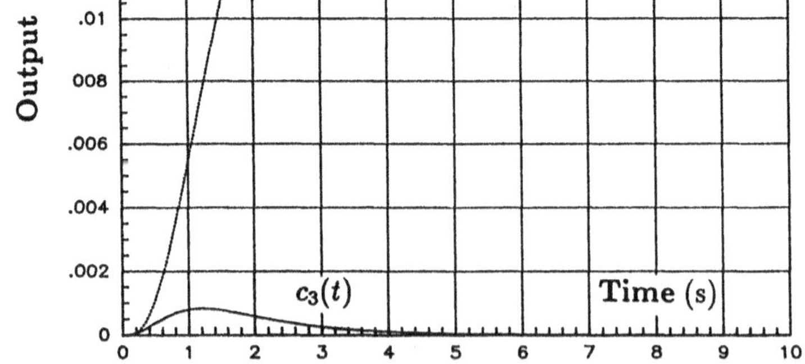

Figure 5.10: L-1011 aircraft: closed–loop responses to a step change in $r_2(t)$.

5.5 The Flight Control System

The final example examines the design of a continuous–time compensator for a fifth–order, three–input, three–output state–space model of a flight control system [75].

The continuous–time state–space matrices of the flight control system are

$$A = \begin{bmatrix} 0 & 0 & 1.132 & 0 & -1.0 \\ 0 & -0.0538 & -0.1712 & 0 & 0.0705 \\ 0 & 0 & 0 & 1.0 & 0 \\ 0 & 0.0485 & 0 & -0.8556 & -1.013 \\ 0 & -0.2909 & 0 & 1.0532 & -0.6859 \end{bmatrix},$$

$$B = \begin{bmatrix} 0 & 0 & 0 \\ -0.0012 & 1.0 & 0 \\ 0 & 0 & 0 \\ 0.4419 & 0 & -1.6646 \\ 0.1575 & 0 & -0.0732 \end{bmatrix},$$

and

$$C = \begin{bmatrix} 1 & 0 & 0 & 0 & 0 \\ 0 & 1 & 0 & 0 & 0 \\ 0 & 0 & 1 & 0 & 0 \end{bmatrix}.$$

This model represents a linearised description of the equations of the motion of an aircraft in the vertical plane when small perturbations about a datum flight condition are considered. The elements of the state vector

$$x(t) = \begin{bmatrix} x_1(t) & x_2(t) & x_3(t) & x_4(t) & x_5(t) \end{bmatrix}^T$$

are as follows:

$x_1(t)$: height error relative to ground or guidance aid (m);

$x_2(t)$: forward speed (m s^{-1});

$x_3(t)$: pitch angle (degree);

$x_4(t)$: rate of change of pitch angle (degree s^{-1}); and

$x_5(t)$: vertical speed (m s^{-1}).

While the elements of the reference input vector

$$r(t) = \left[\begin{array}{ccc} r_1(t) & r_2(t) & r_3(t) \end{array} \right]^T$$

are:

$r_1(t)$: spoiler angle (degree);

$r_2(t)$: forward acceleration due to engine thrust (m s^{-2}); and

$r_3(t)$: elevator angle (degree).

The three state variables, $x_1(t)$, $x_2(t)$ and $x_3(t)$, which are the outputs $c_1(t)$, $c_2(t)$ and $c_3(t)$, respectively, are measurable.

The aim of the design is to ensure that the responses of the three outputs to step changes in their reference inputs are fast, non–oscillatory, non–overshooting, accurate in the steady–state and generally non–interacting.

For the open–loop plant the poles are

$$0, \quad -0.0176 \pm j0.183 \quad \text{and} \quad -0.780 \pm j1.030 \,.$$

For illustrative purposes, it is proposed that the closed–loop system satisfy the following specifications:

- The closed–loop system and the compensator are to be stable.
- All three outputs are to be available for feedback.
- The closed–loop system poles are to be located at -0.2, -0.2, -0.3 and $-1 \pm j0.5$ in the s–plane.
- The steady–state alignment errors of $c_1(t)$ with respect to to the reference input $r_1(t)$, $c_2(t)$ with respect to reference input $r_2(t)$ and $c_3(t)$ with respect to reference input $r_3(t)$ are all zero.
- Steady–state decoupling between the outputs $c_1(t)$, $c_2(t)$ and $c_3(t)$ is to be achieved.

The plant is transformed to its right–coprime RMFD, the numerator N_R and the denominator D_R of which are

$$D_R = \begin{bmatrix} 1.0000 & 0.0000 & 0.0000 & | & 0.8609 & 0.0000 & -0.8365 \\ 0.0000 & 0.0000 & 0.0000 & | & 0.0000 & 1.0000 & 0.0000 \\ 0.0000 & 0.0000 & 1.0000 & | & -0.0945 & 0.0000 & 0.6781 \end{bmatrix} \dots$$

$$\dots \begin{matrix} | & 2.3452 & 2.1224 & -8.7619 \\ | & 0.0437 & 0.0563 & -0.1629 \\ | & 0.3185 & 0.5926 & -1.1794 \end{matrix} \Bigg],$$

degree $\partial[D_R] = 2$, and

$$N_R = \begin{bmatrix} 0.0000 & 0.0000 & 0.0000 & | & -0.1575 & 0.0000 & 0.0732 \\ -0.0012 & 0.0000 & 0.0000 & | & 0.0101 & 1.0000 & -0.0042 \\ 0.0000 & 0.0000 & 0.0000 & | & 0.4419 & 0.0000 & -1.6646 \end{bmatrix}$$

having degree $[N_R] = 1$.

The table below lists the numbers of designer–assigned and solution–assigned closed–loop poles (n_{wd} and n_{ws}, respectively) for a range of feedback compensator orders (n_v). The table shows that a zero–order (static) compensator allows all the five closed–loop poles specified to be designer–assigned (the numbers of designer– and solution–assigned closed–loop poles are estimates only).

n_v	—	n_{ws}	—	n_{wd}
0		0		5

The designer-assigned closed–loop poles are entered and the associated polynomial row block–matrix W_d is constructed, as shown below.

$$W_d = \begin{bmatrix} 0 & 1 & 0 & | & 0 & 2 & 0 & | & 0 & 1.25 & 0 \\ 0 & 0 & 1 & | & 0 & 0 & 0.5 & | & 0 & 0 & 0.06 \\ 0 & 0 & 0 & | & 1 & 0 & 0 & | & 0.2 & 0 & 0 \end{bmatrix},$$

degree $W_d = 2$.

The solution–assigned closed–loop poles and the compensator poles are stable, their positions in the s–plane being $(-7.872,\ -0.4324)$ and $(-7.360,\ -2.044)$, respectively. Thus a second–order feedback compensator is required to assign the five designer–assigned closed–loop poles exactly. The solution–assigned closed–loop poles are located well to the left of the two dominant closed–loop poles, −0.2 and −0.2, and thus their effect on the response of the closed–loop system should be negligible.

The row block–matrices $\tilde{Y}$, $\widehat{Y}$, $\widehat{C}$, $\Gamma_r[\widehat{C}]$, V, K and W_s are calculated. In this example $\tilde{Y}$ and $\widehat{Y}$ are identical.

$$\tilde{Y} = \widehat{Y} = \left[\begin{array}{ccccccccc|c} 0&0&0&0&0&0&0&1&0&1\\ 0&0&0&0&0&1&0&0&0&0\\ 0&0&0&0&1&0&0&0&0&0\\ 0&0&0&1&0&0&0&0&0&0\\ 0&0&1&0&0&0&0&0&0&0\\ 0&1&0&0&0&0&0&0&1&0 \end{array}\right. \cdots$$

$$\cdots \left.\begin{array}{cccccccc} 0.9145 & -0.0434 & 0.0391 & 0.5986 & -0.8709 & 0 & 1.9437 & 0.9145\\ 9.3468 & -0.4346 & 0.4488 & -5.5386 & -6.9653 & 0 & 0 & 9.3468\\ 0.0290 & 0.0084 & -0.0155 & 0.0366 & 0.0692 & 0.0012 & -1 & 0.0290\\ -0.4110 & -0.1190 & 0.2197 & 0.2436 & 0.3151 & 0 & 0 & -0.4110\\ 4.6968 & 0.2230 & -0.8648 & -4.9055 & -8.7866 & 1 & 0 & 4.6968\\ 8.0813 & -0.3411 & 0.4382 & -4.8503 & -6.3375 & 0.0945 & -0.5926 & 7.9032 \end{array}\right],$$

degree $\partial[\tilde{Y}] = 1$. The dimension of the left minimal basis is 6.

$$\widehat{C} = \left[\begin{array}{cccccc|} 0&0&0&0&1&0\\ 0&0&1&0&0&0\\ 0&1&0&0&0&0\\ 1&0&0&0&0&0\\ 0&0&0&0&0&0\\ 0&0&0&0&0&1 \end{array}\right. \cdots$$

$$\cdots \left.\begin{array}{cccccc} 0.0391 & 0.5986 & -0.8709 & 0 & 1.9437 & 0.9145\\ -0.4488 & -5.5386 & -6.9653 & 0 & 0 & 9.3468\\ -0.0155 & 0.0366 & 0.0692 & 0.0012 & -1 & 0.0290\\ 0.2197 & 0.2436 & 0.3151 & 0 & 0 & -0.4110\\ -0.8648 & -4.9055 & -8.7866 & 1 & 0 & 4.6968\\ -0.4382 & -4.8503 & -6.3375 & 0.0945 & -0.5926 & 7.9032 \end{array}\right],$$

degree $\partial[\widehat{C}] = 1$ and

$$\Gamma_r[\widehat{C}] = \left[\begin{array}{ccc|ccc} 0 & 0 & 0 & 0 & 1 & 0\\ 0 & 0 & 1 & 0 & 0 & 0\\ 0 & 1 & 0 & 0 & 0 & 0\\ 1 & 0 & 0 & 0 & 0 & 0\\ -0.8648 & -4.9055 & -8.7866 & 1 & 0 & 4.6968\\ 0 & 0 & 0 & 0 & 0 & 1 \end{array}\right].$$

The row block–matrices V, K and W_s are extracted from rows 1, 5 and 6 of $\widehat{Y}$, yielding:

$$V = \left[\begin{array}{ccc|ccc} 0.0000 & 1.0000 & 0.0000 & 0.0000 & 1.9437 & 0.9145 \\ 0.0000 & 0.0000 & 0.0000 & 1.0000 & 0.0000 & 4.6968 \\ 0.0000 & 0.0000 & 1.0000 & 0.0945 & -0.5926 & 7.9032 \end{array}\right],$$

degree $\partial[V] = 1$;

$$K = \begin{bmatrix} 0.0391 & 0.5986 & -0.8709 \\ -0.8648 & -4.9055 & -8.7866 \\ -0.4382 & -4.8503 & -6.3375 \end{bmatrix},$$

degree $\partial[K] = 0$; and

$$W_s = \left[\begin{array}{ccc|ccc} 0.0000 & 0.0000 & 0.0000 & 1.0000 & 0.9145 & -0.0434 \\ 0.0000 & 0.0000 & 1.0000 & 0.0000 & 4.6968 & 0.2230 \\ 0.0000 & 1.0000 & 0.0000 & 0.0000 & 8.0813 & -0.3411 \end{array}\right],$$

degree $\partial[W_s] = 1$.

The steady–state specifications are next satisfied. The row block–matrix M, the numerator of the pre–compensator $R(s)$, is found to be

$$M = \begin{bmatrix} 0.0391 & 1.2500 & -0.0344 \\ -0.8648 & 0.0000 & -0.2073 \\ -0.4382 & 0.0000 & -0.3106 \end{bmatrix},$$

degree $\partial[M] = 0$.

The behaviour of the closed–loop system to unit step changes applied to the first, second and third inputs in turn, is shown in Fig. 5.11, 5.12 and 5.13, respectively. The flight control system (i) exhibits an improved speed of response which is non–oscillatory and non–overshooting, (ii) satisfies steady–state error and decoupling requirements, and (iii) reduces interaction under transient conditions.

The LMFD of the compensator is

$$F(s) = \begin{bmatrix} 0.0000 & s + 1.9437 & 0.9145 \\ 1.0000 & 0.0000 & 4.6968 \\ 0.0945 & -0.5926 & s + 7.9032 \end{bmatrix}^{-1} \cdot \begin{bmatrix} 0.0391 & 0.5986 & -0.8709 \\ -0.8648 & -4.9055 & -8.7866 \\ -0.4382 & -4.8503 & -6.3375 \end{bmatrix}$$

and

$$R(s) = \begin{bmatrix} 0.0000 & s + 1.9437 & 0.9145 \\ 1.0000 & 0.0000 & 4.6968 \\ 0.0945 & -0.5926 & s + 7.9032 \end{bmatrix}^{-1} \cdot \begin{bmatrix} 0.0391 & 1.2500 & -0.0344 \\ -0.8648 & 0.0000 & -0.2073 \\ -0.4382 & 0.0000 & -0.3106 \end{bmatrix}.$$

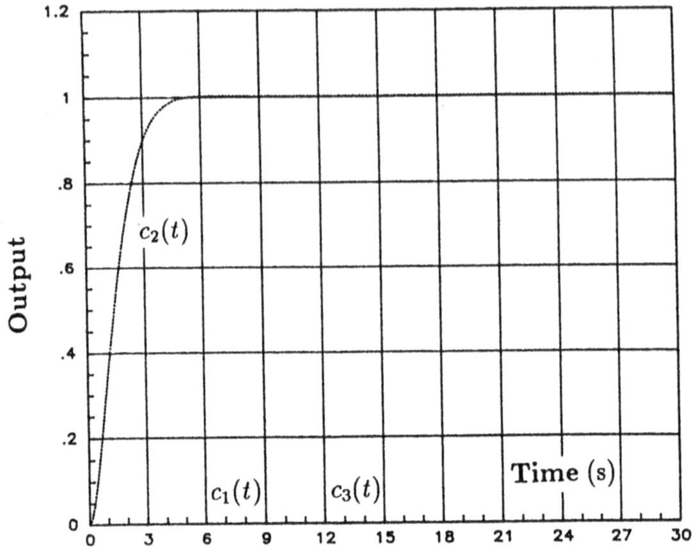

Figure 5.11: Flight control: closed–loop responses to a step change in $r_1(t)$.

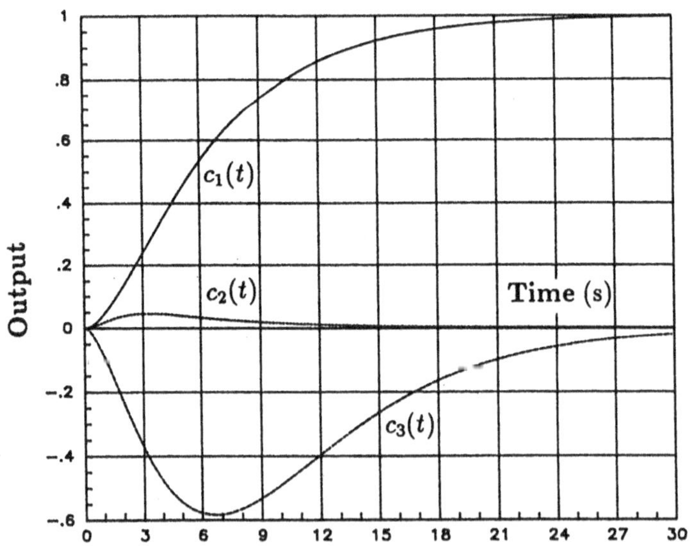

Figure 5.12: Flight control: closed–loop responses to a step change in $r_2(t)$.

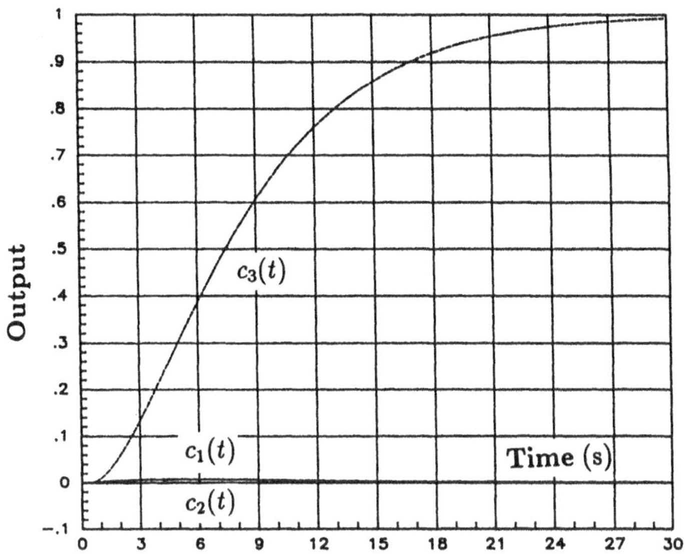

Figure 5.13: Flight control: closed–loop responses to a step change in $r_3(t)$.

Alternatively, the compensator can be represented by its SSD. The SSD of the feedback compensator is

$$A = \begin{bmatrix} -1.8188 & -4.0731 \\ 0.3059 & -7.5843 \end{bmatrix}, \ B = \begin{bmatrix} 0.0045 & 0.0405 & 0.2525 \\ 0.0229 & 0.2815 & 0.3534 \end{bmatrix},$$

$$C = \begin{bmatrix} 0.0000 & 73.1886 \\ -8.0445 & 3.2824 \\ 0.0000 & -15.5828 \end{bmatrix}, \ E = \begin{bmatrix} -0.8648 & -4.9055 & -8.7866 \\ 0.0000 & 0.0000 & 0.0000 \\ 0.000 & 0.000 & 0.0000 \end{bmatrix},$$

where E is the transmission matrix. Likewise, the SSD of the pre–compensator is

$$A = \begin{bmatrix} -7.3927 & -1.7709 \\ 0.1008 & -2.0104 \end{bmatrix}, \ B = \begin{bmatrix} 0.3459 & 0.0878 & 0.2777 \\ 0.0041 & 0.1326 & -0.0036 \end{bmatrix},$$

$$C = \begin{bmatrix} 4.8791 & -3.2297 \\ 0.0000 & 9.4281 \\ -1.0388 & 0.6877 \end{bmatrix}, \ E = \begin{bmatrix} -0.8648 & 0.0000 & -0.2073 \\ 0.0000 & 0.0000 & 0.0000 \\ 0.000 & 0.000 & 0.0000 \end{bmatrix}.$$

The eigenvalues of the A matrix of the compensator are those calculated earlier, i.e., -7.360 and -2.044. The SSDs of the compensator are available to the designer.

Chapter 6

Conclusions and Suggestions for Future Work

An aim of this work was to devise a direct method for the design of *dynamic* compensators for MIMO control systems. To this end a new class of computational techniques and an associated methodology, based on algebraic methods, have been proposed and developed for the design of dynamic compensators, be they analog or digital. The computational tools have been implemented in a menu–driven design software package. The design of the dynamic performance of the closed–loop system is based on pole assignment; the steady–state performance is framed in terms of alignment and following errors.

Techniques have been developed for:

- the computation of the determinant and inverse of a non–singular polynomial matrix, the reduction of a full–rank polynomial matrix to its row–proper form, the construction of a non–singular polynomial matrix the eigenvalues of which are specified by the designer;
- the transformations between the various models of a linear system, namely, state–space to matrix fraction description, and matrix fraction description to transfer function matrix description.
- designing a dynamic two–element compensator for a MIMO closed–loop control system such that steady–state and dynamic performance specifications are realised.

To achieve the above results, a number of techniques was developed some of which are listed below:

1. The formation of the resultant of a polynomial matrix, and its application to the multiplication of polynomial matrices.

2. The concept of the "structure" matrix and its numerical representation known as the "skeleton" block-matrix.

3. A methodology to transform a polynomial matrix to its regular form. This methodology is used not only to realise the ARMA model of a matrix fraction description, the denominator of which is not regular, but also to calculate the determinant of a polynomial matrix.

4. A methodology to transform a SSD to its coprime MFD.

5. The numerical representation of the modified left compensator equation, using the method of resultants.

6. The problem of solving the modified left compensator equation to realise a coprime, proper, least-order output-feedback compensator. The problem is cast as a minimal design problem.

7. The application of a numerical method described by Kung and Kailath [58] to solve the minimal design problem.

8. A methodology for the design of a pre-compensator which satisfies the steady-state alignment- or following-error specifications.

The advantage of the design methodology is that it enables the designer to carry out a number of trial designs. These designs can be tested to ascertain sensitivities to parameter variations, etc., and modified. The software package, in which the design techniques are implemented, allows the user to exit from it into a commercial host package [53] which possesses some powerful facilities for the analysis and simulation of MIMO control systems.

In addition, the user has access to the variables and their values, scalar or matrix, in the host software. With these results as a starting point, further investigations can be conducted using the features of the host software.

The design of the dynamics of the closed–loop system employs pole–assignment techniques which suffer from a number of disadvantages. Further work is required not only to overcome the deficiencies but also to investigate the possible use of other design methods which employ and extend some of the techniques developed in the earlier chapters. Using state–space techniques, research into methods for improving pole assignment – or eigenstructure assignment – by output feedback continues [36] [116]. Although the designs may result in static compensators, such approaches may provide ideas for application to the method discussed in Chap. 4.

Even if the poles of the closed–loop system lie in desired locations, the outputs may not respond in the desired manner to test signals applied to the reference inputs. Assignment of both the poles and *zeros* of the closed–loop system would enable the response to be specified completely. Though based on state–space techniques [76] [11], progress on this front may provide some inspiration for application to techniques based on algebraic methods.

In the design methodology presented in the earlier chapters, there is no procedure for selecting the output–feedback compensator out of the infinite set which best satisfies the pole–assignment specifications and other design requirements simultaneously. This problem is aggravated by the fact that the designer–assigned polynomial matrix, $W_{\mathrm{d}}(q)$, is not unique for a selected set of eigenvalues. (The order in which these are entered into the software may yield an alternative but valid polynomial matrix.) Further work is required to determine how the extra degrees of freedom can be employed to satisfy some of the other design constraints or specifications, and hence yield a more effective compensator design.

A method for specifying the steady–state performance of the MIMO system in terms of alignment or following errors has been presented. An advantage of the approach is that decoupling of the steady–state errors results – and no doubt introduces some decoupling during the transient response. However, the steady-state errors and accuracy may be sensitive to parameter variations in the closed-loop system. Because of this approach for determining steady-state performance, and the use of pole assignment for the design of dynamic performance, the advantages of the use of high loop gains in conventional design procedures are lost. High loop gain not only improves steady-state accuracy but also reduces interaction between loops, sensitivity to external disturbances and sensitivity to parameter variations.

In essence, there is a need to provide a facility for designing *dynamic* compensators, perhaps using an approach similar to that described in Chap. 4, but which includes some of the desirable features of other design approaches. How this is achieved – by reformulating the problem, altering the structure of the compensator, or by utilizing the extra degrees of freedom – is a challenge for future research.

Appendix A

Proof: The Numerical Block–Matrix Multiplication of Two Conformable Polynomial Matrices

This appendix contains the proof of Theorem 2.3.

Proof A.1 The inductive proof [45] of Theorem 2.3 is based upon two observations. Firstly, expressing the multiplicand in the form

$$C(q) = q^{\partial[C]}.C_0 + C_{\mathrm{R}}(q), \tag{A.1}$$

we note that eqn. (2.7) can be written as

$$E(q) = q^{\partial[C]}.C_0.L(q) + C_{\mathrm{R}}(q).L(q)\,. \tag{A.2}$$

Secondly, the resultant $< L >_{\partial[C]}$ may be partitioned as follows

$$< L >_{\partial[C]} = \begin{bmatrix} L & 0_{\partial[C]} \\ 0^b_{\partial[C]} & < L >_{(\partial[C]-1)} \end{bmatrix}, \tag{A.3}$$

where $0_{\partial[C]}$ denotes the $\partial[C]$–element, null row–block matrix.
Equation (2.8),

$$E = C. < L >_{\partial[C]}, \tag{A.4}$$

may be expanded to the form

$$E = C_0.\begin{bmatrix} L & 0_{\partial[C]} \end{bmatrix} + \begin{bmatrix} 0 & [C_1 \ldots C_{\partial[C]}]. < L >_{(\partial[C]-1)} \end{bmatrix}. \tag{A.5}$$

Comparing eqn. (A.5) with (A.2), we note that $C_0.[L \;\; 0_{\partial[C]}]$ represents the term $q^{\partial[C]}.C_0.L(q)$ and

$$\left[\, C_1 \;\; \dots \;\; C_{\partial[C]} \,\right] . < L >_{(\partial[C]-1)}$$

the product $C_{\mathrm{R}}(q).L(q)$. Repeated expansions of the terms of the form $C_{\mathrm{R}}(q).L(q)$ will complete the proof.

Similarly, eqn. (2.9) can be proven.

Appendix B

The Lower–Triangular Toeplitz Block–Matrix and its Inverse

B.1 Introduction

Toeplitz or Hankel block–matrices have the property that their diagonals or anti–diagonals contain block–matrices of identical coefficients. In this appendix the structure of the lower–triangular Toeplitz block–matrix and a simple iterative technique for computing its inverse [5,21,45] are discussed.

Consider a $n \times p$ polynomial matrix $L(q)$ of degree $\partial[L(q)] = m$, for which the column block–matrix description is

$$L^b = \begin{bmatrix} L_0 \\ L_1 \\ \vdots \\ L_m \end{bmatrix} .$$

Let us consider the structure of the lower–triangular Toeplitz block–matrix associated with the polynomial matrix $L(q)$.

The lower–triangular Toeplitz block–matrix of *order* g is denoted by $TL_g(L^b)$. Let $g \geq m$, $h = g - m$, then

$$TL_g(L^b) = TL_g([L_0\, L_1\, \dots\, L_m]^b)$$

$$= \begin{bmatrix} L_0 & & 0 & \dots & \dots & 0 \\ \vdots & \ddots & \vdots & & & \vdots \\ \vdots & L_0 & 0 & \dots & \dots & 0 \\ \vdots & \vdots & L_0 & & & \\ L_m & \vdots & \vdots & \ddots & & \\ \ddots & \vdots & \vdots & & \ddots & \\ & L_m & L_{m-1} & \dots & \dots & L_0 \end{bmatrix} \tag{B.1}$$

$$= \left[\begin{array}{c|c} & 0 \\ <L>^b_h & \text{------} \\ & TL_{m-1}([L_0 \dots L_{m-1}]^b) \end{array} \right].$$

Remark B.1 The lower–triangular Toeplitz block–matrix, $TL_g(L^b)$, of order g consists of $(g+1) \times (g+1)$ block–matrices, L_i, of dimension $n \times p$ and is nonsingular if and only if L_0 is nonsingular because

$$det\{TL_g(L^b)\} = \{det(L_0)\}^{g+1} .$$

With reference to eqn. (B.1) the lower–triangular Toeplitz block–matrix can be separated into two sub–matrices when the order g of the lower–triangular Toeplitz block–matrix is greater than the degree m of $L(q)$. The Toeplitz block–matrix can then be constructed by computing the resultant, $<L>^b_h$, of order h and the lower–triangular block–matrix, $TL_{m-1}([L_0\, \dots\, L_{m-1}]^b)$, of reduced order $(m-1)$ of L^b. When the order g of the lower–triangular Toeplitz block–matrix is less than or equal to the degree m of $L(q)$, the general form of the lower–triangular block–matrix, i.e. eqn. (B.1), takes on a simplified form. The resultant, $<L>^b_h$, of order h found in the general form disappears; the lower–triangular Toeplitz block–matrix is then of the form

$$TL_g(L^b) = TL_g([L_0\, L_1\, \dots\, L_m]^b)$$

$$= \begin{bmatrix} L_0 & 0 & \dots & \dots & 0 \\ \vdots & L_0 & & \ddots & \vdots \\ \vdots & \vdots & \ddots & \ddots & \vdots \\ \vdots & \vdots & & \ddots & 0 \\ \vdots & \vdots & & & L_0 \\ \vdots & \vdots & & & \vdots \\ L_m & L_{m-1} & \dots & \dots & L_{m-g} \end{bmatrix}, \tag{B.2}$$

for $g \leq m$.

B.2 The Inverse of a Lower–Triangular Toeplitz Block–Matrix

Let

$$L^b = \begin{bmatrix} L_0 & \dots & L_m \end{bmatrix}^b$$

denote the column block–matrix of the polynomial matrix $L(q)$ of degree m, and let

$$E^b = \begin{bmatrix} E_0 & \dots & E_g \end{bmatrix}^b$$

denote the column block–matrix of the polynomial matrix $E(q)$ of degree g.

The inverse of a lower–triangular Toeplitz block–matrix is also a lower–triangular Toeplitz block–matrix. Based on this result, the inversion algorithm [5,21,45,54,127] can be derived as follows.

Let us determine the inverse of the lower–triangular Toeplitz block–matrix, $TL_g^{-1}(L^b)$, of order g such that

$$TL_g^{-1}(L^b) = TL_g(E^b)\,.$$

Hence

$$TL_g(E^b).TL_g(L^b) = TL_g(L^b).TL_g(E^b) = I$$

or

$$\begin{bmatrix} E_0 & 0 & \dots & \dots & 0 \\ \vdots & \ddots & \ddots & & \vdots \\ \vdots & & \ddots & \ddots & \vdots \\ \vdots & & & \ddots & \vdots \\ E_g & \dots & \dots & \dots & E_0 \end{bmatrix} . \begin{bmatrix} L_0 & 0 & \dots & \dots & 0 \\ \vdots & \ddots & \ddots & & \vdots \\ L_m & & \ddots & \ddots & \vdots \\ & \ddots & & \ddots & 0 \\ & & L_m & \dots & L_0 \end{bmatrix}$$

$$= \begin{bmatrix} L_0 & 0 & \dots & \dots & 0 \\ \vdots & \ddots & \ddots & & \vdots \\ L_m & & \ddots & \ddots & \vdots \\ & \ddots & & \ddots & 0 \\ & & L_m & \dots & L_0 \end{bmatrix} . \begin{bmatrix} E_0 & 0 & \dots & \dots & 0 \\ \vdots & \ddots & \ddots & & \vdots \\ \vdots & & \ddots & \ddots & \vdots \\ \vdots & & & \ddots & \vdots \\ E_g & \dots & \dots & \dots & E_0 \end{bmatrix}$$

$= I$. Thus

$$E_0.L_0 = L_0.E_0 = I,$$

and, for $0 < k \le g$,

$$\begin{bmatrix} E_k & \dots & E_0 \end{bmatrix} . \begin{bmatrix} L_0 \\ \vdots \\ L_k \end{bmatrix} = 0, \text{ or } \begin{bmatrix} L_k & \dots & L_0 \end{bmatrix} . \begin{bmatrix} E_0 \\ \vdots \\ E_k \end{bmatrix} = 0 .$$

Hence

$$E_0 = L_0^{-1} \tag{B.3}$$

and, for $k = 1, 2, \dots, g$,

$$\begin{aligned} E_k &= -\begin{bmatrix} E_{k-1} \dots E_0 \end{bmatrix} . \begin{bmatrix} L_1 \\ \vdots \\ L_k \end{bmatrix} .L_0^{-1} \\ &= -L_0^{-1}. \begin{bmatrix} L_k & \dots & L_1 \end{bmatrix} . \begin{bmatrix} E_0 \\ \vdots \\ E_{k-1} \end{bmatrix} . \end{aligned} \tag{B.4}$$

It is assumed for $k > m$, $L_k = 0$.

The inversion algorithm for the lower-triangular Toeplitz block-matrix is extremely simple for the special case of a first degree, monic polynomial matrix

$$L(q) = q.I - A \text{ or } L^b = \begin{bmatrix} I \\ A \end{bmatrix}$$

then, from eqns. (B.3) and (B.4)

$$E = \begin{bmatrix} I \\ A \\ A^2 \\ \vdots \\ A^g \end{bmatrix} .$$

Therefore

$$TL_g^{-1}\left(\begin{bmatrix} I \\ -A \end{bmatrix}\right) = TL_g\left(\begin{bmatrix} I \\ A \\ \vdots \\ A^g \end{bmatrix}\right) .$$

There exists a dual description of the lower-triangular Toeplitz block-matrix known as the upper-triangular Toeplitz block-matrix. The upper-triangular Toeplitz block-matrix has similar properties to that of the lower-triangular Toeplitz block-matrix [45].

Appendix C

Algorithm to Transform a Polynomial Matrix into a Regular Form

C.1 Introduction

It has been shown [90] that a non-singular polynomial matrix $L(q)$ can always be transformed to a regular or monic form by multiplication by a nilpotent polynomial matrix $T(q)$ (see Definition C.1) such that $\det(T(q)) = c.q^k$. The algorithm for the calculation of $T(q)$ is based on real-matrix factorisation techniques hence satisfactory numerical properties are assured.

The methodology is applied to two types of transformation, $T1$ and $T2$, which are of special interest.

$T1$ Consider a LMFD $(\widehat{D}_{\mathrm{L}}^{-1}(q).\widehat{N}_{\mathrm{L}}(q))$, the denominator polynomial matrix, $\widehat{D}_{\mathrm{L}}(q)$, of which is not regular. A polynomial matrix $T(q)$ is sought such that by pre-multiplying the numerator and denominator by $T(q)$; i.e., $N_{\mathrm{L}}(q) = T(q).\widehat{N}_{\mathrm{R}}(q)$ and $D_{\mathrm{L}}(q) = T(q).\widehat{D}_{\mathrm{L}}(q)$, the denominator polynomial matrix is transformed to a regular polynomial matrix. This transformation is of particular interest when the difference equation (ARMA model) of a discrete-time system is sought (refer to § 3.6).

$T2$ Consider a square non-singular polynomial matrix $L(q)$ with a *singular leading* coefficient block-matrix; i.e., $L(q)$ is not regular. A polynomial matrix $T(q)$ is sought such that by post–multiplying the polynomial matrix $L(q)$ by $T(q)$ the polynomial matrix $L(q).T(q)$ is transformed to a regular polynomial matrix, where $\det(L(q).T(q)) = \det(L(q)).c.q^k$ (refer to § 2.4.8).

Definition C.1 : The polynomial matrix $T(q)$ is called *nilpotent* if all its eigenvalues are zero, i.e., if $\det(T(q)) = c.q^k$, where c and k are constants.

Proposition C.1 Let

$$L(q) = L_0 q^m + L_1 q^{m-1} + \cdots + L_m$$

denote a $n \times n$ non–singular polynomial matrix of degree m and let

$$\det(L(q)) = l(q) = l_0 q^g + l_1 q^{g-1} + \cdots + l_g$$

denote the determinant polynomial of $L(q)$ of polynomial order g.

The polynomial order, g, of the determinant of the matrix $L(q)$ is related to its dimension, n, and degree, m, by the following inequality (see Result 2.2):

$$g \leq m.n\,. \tag{C.1}$$

The equality holds if and only if $L(q)$ is regular. In other words, the *leading coefficient*, L_0, is non–singular if and only if

$$g = n.m\,. \tag{C.2}$$

In order to transform a non–singular polynomial matrix with the *singular leading* coefficient, L_0, into a *regular* form we have to, according to eqn. (C.1), increase the number of eigenvalues by

$$k = m.n - g \tag{C.3}$$

maintaining the degree m. It can be accomplished by multiplication of $L(q)$ by a properly chosen $T(q)$ as follows:

$$L_{\mathrm{R}}(q) = L(q).T(q), \text{ or } L_{\mathrm{R}}(q) = T(q).L(q)\,. \tag{C.4}$$

The polynomial matrix $T(q)$ can be chosen such that:

- $L_{\mathrm{R}}(q)$ is a regular polynomial matrix of degree equal to the degree of $L(q)$, which is m, thus, having exactly $m.n$ eigenvalues; and

- $T(q)$ has $k = m.n - g$ eigenvalues.

Furthermore, $T(q)$ can be chosen to be nilpotent, i.e., all its k eigenvalues are zero. Equation (C.4) describes the desired transformation of $L(q)$ into a regular polynomial matrix $L_{\mathrm{R}}(q)$. [90].

Definition C.2 Let us consider a $n \times n$ first degree polynomial matrix $U(q)$ which can be thought of as a canonical form of elementary nilpotent polynomial matrices,

$$U(q) = U_0 q + U_1 = \begin{bmatrix} 0 & qI_k \\ I_r & 0 \end{bmatrix}, \tag{C.5}$$

where the block-matrices U_0 and U_1 are $n \times n$ matrices, I_n is the $n \times n$ identity matrix and $n = k + r$.

The row block-matrix description of $U(q)$ is of the form

$$U = \begin{bmatrix} U_0 & U_1 \end{bmatrix} = \begin{bmatrix} 0_r & I_n & 0_k \end{bmatrix}, \tag{C.6}$$

where 0_k is k–column matrix of zeros.

The determinant of $U(q)$ is given by

$$\det(U(q)) = \pm q^k \,. \tag{C.7}$$

Remark C.1 It can be shown that post-multiplication of the polynomial matrix $L(q)$ by $U(q)$ left-shifts the row block-matrix of coefficients $L = [L_0 \ldots L_m]$ by k columns.

Because of this shifting property the nilpotent matrix $U(q)$, defined in eqn. (C.5), is called the *column-shift polynomial matrix* (c.s.p.m.) of order k.

In § C.3 the row-shift polynomial matrix, which is the dual to that of eqn. (C.5), is considered.

C.2 Transformation of a Polynomial Matrix into a Regular Form by Post–Multiplication

In order to perform the transformation described by eqn. (C.4), the nilpotent polynomial matrix $T(q)$ must be calculated; an algorithm is presented in this section.

The algorithm searches for n linearly independent columns in the row block–matrix $L = [L_0 \ldots L_m]$ by a succession of appropriate matrix decompositions. The linearly independent columns are relocated to the left–hand side of the row block–matrix L. The matrices that describe the sequence of the column relocations form the polynomial matrix $T(q)$.

Theorem C.1 : Let $L(q)$ be a $n \times n$ nonsingular polynomial matrix of degree m as in Proposition C.1. There exists a nilpotent polynomial matrix $T(q)$ such that:

1. the product $L_{\mathrm{R}}(q) = L(q).T(q)$ is regular; and

2. $T(q)$ is the product of the first degree, $n \times n$ nilpotent polynomial matrices

$$T(q) = T_{(p)}(q) = T^{(0)}(q).T^{(1)}(q).T^{(2)}(q)\ldots T^{(p-1)}(q).T^{(p)}(q),\ 1 \le p \le m \qquad \text{(C.8)}$$

such that

$$T^{(i)}(q) = Q^{(i)}.U^{(i)}(q), \qquad \text{(C.9)}$$

where:

$Q^{(i)}$ is a $n \times n$ non–singular constant matrix; and

$U^{(i)}(q)$ is the column shift polynomial matrix of order k_i, $0 \le k_i \le n$ (see Definition C.2).

Result C.1 The determinant of $T(q)$ is

$$\det(T(q)) = \prod_{i=1}^{p} \det(Q^{(i)}.U^{(i)}(q)) = cq^k,$$

where

$$c = \prod_{i=1}^{p} \det(Q^{(i)})$$

and

$$k = \partial[\det(T(q))] = \sum_{i=1}^{p} k_i \,.$$

Algorithm C.1 Let us now consider the recursive algorithm which transforms $L^{(0)}(q) \triangleq L(q)$ into a regular polynomial matrix $L_{\mathrm{R}}(q) \triangleq L^{(p)}(q)$.

Set $i = 0$, $L^{(0)}(q) = L(q)$ and $T^{(0)}(q) = I_n$ to start the algorithm.

Consider the ith iteration in the evaluation of $T(q)$ based on eqn. (C.8).

step 1. If $r_i = rank(L_0^{(i-1)}) = n$, the algorithm terminates and the polynomial matrix $T(q) = T_{(i-1)}(q)$, set $p = i - 1$;

if $r_i < n$, factorise $L_0^{(i-1)}$ (say, using QR, LDU or SVD matrix decomposition) into

$$L_0^{(i-1)} = \begin{pmatrix} 0_{k_i} & L_{0D}^{(i)} \end{pmatrix} .(Q^{(i)})^{-1};\ \text{i.e., } L_0^{(i-1)}.Q^{(i)} = \begin{pmatrix} 0_{k_i} & L_{0D}^{(i)} \end{pmatrix}, \tag{C.10}$$

where $Q^{(i)}$ is a $n \times n$ non–singular (usually unitary) matrix, $k_i = n - r_i$ and 0_{k_i} is a k_i–column zero matrix.

step 2. Postmultiply $L^{(i-1)}(q)$ by matrix $Q^{(i)}$

$$\hat{L}(q) = L^{(i-1)}(q).Q^{(i)}, \tag{C.11}$$

(the leading coefficient of $\hat{L}(q)$ is now equal to the right hand side of eqn. (C.10)).

step 3. Postmultiply $\hat{L}(q)$ by the c.s.p.m of order k_i

$$L^{(i)}(q) = \hat{L}(q).U^{(i)}(q), \tag{C.12}$$

(this multiplication shifts the matrix of coefficients of $\hat{L}(q)$, to the left by k_i columns of zeros). Update the transformation matrix

$$T_{(i)}(q) = T_{(i-1)}(q).T^{(i)}(q)\,. \tag{C.13}$$

This ends the ith iteration. Combining eqns. (C.10) to (C.13), the ith iteration of the algorithm results in

$$L^{(i)}(q) = L^{(i-1)}(q).Q^{(i)}.U^{(i)}(q) = L^{(i-1)}(q).T^{(i)}(q) = L(q).T_{(i)}(q), \tag{C.14}$$

where $T^{(i)}(q)$ and $T_{(i)}(q)$ are defined by eqns. (C.9) and (C.13). The final iteration $(p = i - 1)$ yields

$$L_{\mathrm{R}}(q) = L^{(p)}(q) = L(q).(T(q) = T_{(p)}(q)),$$

where $T(q)$ is defined by eqn. (C.8).

Comment: Algorithm C.1 can be used to implement the transformation type $T2$.

C.3 Transformation of a Polynomial Matrix into a Regular Form by Pre–Multiplication

The transformation of a square polynomial matrix, $L(q)$, into its regular form can be accomplished by means of pre–multiplication by an appropriately chosen polynomial matrix $T(q)$, shown below,

$$L^{(i)}(q) = T^{(i)}(q).L^{(i-1)}(q) = U^{(i)}(q).Q^{(i)}.L^{(i-1)}(q)\,. \tag{C.15}$$

This equation should be compared with eqn. (C.14).

In this case it is convenient to consider the column block–matrix description, L^b, of $L(q)$; i.e.,

$$L^b = \begin{bmatrix} L_0 \\ \vdots \\ L_m \end{bmatrix}.$$

The matrix $Q^{(i)}$ and $r_i = rank(L_0^{(i-1)})$ are now defined by the following decomposition

$$L_0^{(i-1)} = (Q^{(i)})^{-1}.\begin{bmatrix} 0_{k_i} \\ L_{0D} \end{bmatrix}, \tag{C.16}$$

where 0_{k_i} denotes a matrix with $k_i = n - r_i$ zero rows.

It is also convenient to consider a row–shift polynomial matrix, $U(q)$, with the matrix of coefficients arranged in a column block–matrix as follows

$$U(q) = \begin{bmatrix} 0 & I_r \\ qI_k & 0 \end{bmatrix},$$

the column block–matrix description of which is

$$U^b = \begin{bmatrix} U_0 \\ U_1 \end{bmatrix} = \begin{bmatrix} 0_r \\ I_n \\ 0_k \end{bmatrix},$$

where 0_k denotes a k–row matrix of zeros.

Pre–multiplication by $U(q)$ shifts a matrix by k rows upwards. This pre–multiplication should be compared with the post–multiplication in eqn. (C.14).

Comment: An algorithm similar to that of Algorithm C.1, but with the necessary modifications, can be used to implement the transformation type $T1$.

Appendix D

Pseudo–Inverse of a Resultant

The pseudo–inverse of a resultant is used by Algorithm 3.2 of § 3.4 to compute the inverse of a polynomial matrix.

The resultant is a rectangular matrix and therefore an inverse does not exist, however, it is possible to define various pseudo–inverses. Let us consider the general description of the pseudo–inverse of a resultant.

Let

$$L = \begin{bmatrix} L_0 & L_1 & \dots & L_m \end{bmatrix}$$

denote the row block–matrix of coefficients of $L(q)$ of degree $\partial[L(q)] = m$, and dimension n. Let $< L >_h$ denote the resultant of order h. The pseudo–inverse $Z_k^{-1}(< L >_h)$ of the resultant is any matrix which satisfies

$$< L >_h . Z_k^{-1}(< L >_h) = \begin{bmatrix} 0_{n.k} & I_{n.(h+1)} & 0_{n.(m-k)} \end{bmatrix}, \qquad \text{(D.1)}$$

where the subscript k of $Z_k^{-1}(< L >_h)$ denotes the *order* of the pseudo–inverse, $0_{n.k}$ denotes a $n.k$ column matrix of zeros, $I_{n.(h+1)}$ is the $n.(h+1) \times n.(h+1)$ identity matrix, and $0_{n.(m-k)}$ denotes a $n.(m-k)$ column matrix of zeros.

Postmultiplying eqn. (D.1) by the matrix $Z_k(< L >_h)$, which must be non–singular, we have

$$< L >_h = \begin{bmatrix} 0_{n.k} & I_{n.(h+1)} & 0_{n.(m-k)} \end{bmatrix} . Z_k(< L >_h) . \qquad \text{(D.2)}$$

In order for eqns. (D.1) and (D.2) to be satisfied it is sufficient for the matrix $Z_k(< L >_h)$

to be of the form

$$Z_k(< L >_h) = \begin{bmatrix} A \\ < L >_h \\ B \end{bmatrix}, \tag{D.3}$$

where $k \in \{0,\ 1,\ \ldots\}$, $0 \leq k \leq m$, and the matrices A and B, of dimensions $n.k \times n.(m+h+1)$ and $n.(m-k) \times n.(m+h+1)$, respectively, are chosen such that the matrix $Z_k(< L >_h)$ is square, non-singular and of dimension $n.(m+h+1) \times n.(m+h+1)$.

Consider the case when the order of the pseudo-inverse is $k = 0$ and the polynomial matrix $L(q)$ is square and non-singular. Thus, with reference to Theorem 2.2, the resultant $< L >_h$ is of full row rank. Then

$$< L >_h .Z_0^{-1}(< L >_h) = \begin{bmatrix} I_{n.(h+1)} & 0_{n.m} \end{bmatrix},$$

where

$$Z_0(< L >_h) = \begin{bmatrix} < L >_h \\ B \end{bmatrix}.$$

It is sufficient that $n.m \times n.(h+m+1)$ matrix B consist of $n.m$ linearly independent rows of, say, the $n.(h+m+1) \times n.(h+m+1)$ identity matrix, for $Z_0(< L >_h)$ to be square and non-singular. The $n.m$ rows of the identity matrix, which constitute the matrix B, are chosen initially by determining which columns of $< L >_h$ are linearly dependent. This is accomplished by systematically searching through the columns of $< L >_h$, beginning with its leftmost column, until it is found, say, that the jth column of $< L >_h$ is linearly dependent with respect to the previous columns of $< L >_h$ that are linearly independent. Then the jth row of the $n.(m+h+1) \times n.(m+h+1)$ identity matrix is a row (any row) of the matrix B. This technique generates a square non-singular matrix $Z_0(< L >_h)$.

If, say, the polynomial matrix $L(q)$ is regular, that is, L_0 is of rank n, then the first $n.(h+1)$ columns of its resultant of order h are linearly independent. Therefore, the matrix B is composed of the bottom $n.m$ rows of the $n.(m+h+1) \times n.(m+h+1)$ identity matrix, hence

$$Z_0(< L >_h) = \begin{bmatrix} < L >_h \\ [0_{m.n \times n.(h+1)} \ I_{m.n}] \end{bmatrix}.$$

Appendix E

Calculation of the Controllability Indices of a State–Space System

The following methodology found in Kailath [57] is used to realise the set of controllability indices $I_r(A,B) \equiv \{\mu_1, \ldots, \mu_{n_i}\}$ of a completely controllable state–space system $S(A,B,C,E)$.

It will be convenient, following Kalman [59], to use a so–called *crate* diagram as an aid to describing the following selection procedure used for realising the controllability indices of $S(A,B,C,E)$.

The crate, shown in Fig. E, is a table with n_i columns, representing the columns of the input matrix $B = [b_1 b_2 \ldots b_{n_i}]$, and n_s rows corresponding to the powers $\{I, A, \ldots, A^{n_s-1}\}$. The (i,j)th cell of the crate then represents the column vector $A^{i-1}.b_j$. Choosing n_s linearly independent columns of

$$CT_{n_s-1} = \left[\begin{array}{cccc} B & A.B & \ldots & A^{n_s-1}.B \end{array} \right]$$

corresponds to selecting n_s cells of the crate. Clearly this can be done in many ways, but we shall consider one procedure which is known as Scheme II (search the crate by rows) [57].

We first select b_1 and indicate this by putting an 'x' in the $A^0.b_1$ cell. Now if b_2 is linearly independent of b_1, we enter 'x' in this cell as well and continue along the first row of the crate until we either place 'x' in all the cells or we find a vector, say $A^{k_i}.b_i$, that is linearly dependent on *all* the previously selected vectors. A '0' is inserted in the corresponding cell; all the remaining vectors below it in the same column will also be linearly dependent on the already–selected vectors. Therefore all the associated cells are left blank and we proceed, if necessary, to the next linearly independent vector encountered in the row search. (Searching the crate by rows corresponds to examining the columns of the controllability matrix from left to right.)

We now have selected a subset of n_s linearly independent vectors

$$b_1,\ A.b_1,\ \ldots,\ A^{\mu_1 - 1}.b_1,\ b_2,\ A.b_2,\ \ldots,\ A^{\mu_{n_i} - 1}.b_{n_i} \tag{E.1}$$

of $CT_{n_s - 1}$. The set of indices $\{\mu_1, \mu_2, \ldots, \mu_{n_i}\}$ represent the set of controllability indices, $I_r(A, B)$, of $S(A, B, C, E)$.

A typical crate diagram produced by this scheme is shown in Fig. E. In the latter the controllability indices are $\mu_1 = 3$, $\mu_2 = 1$, $\mu_3 = 3$ and $\mu_4 = 0$ for $n_s = 7$ and $n_i = 4$.

b_1	b_2	b_3	b_4	
x	x	x	0	I
x	0	x		A
x		x		A^2
0		0		A^3
				A^4
				A^5
				A^6

Figure E.1: Typical crate diagram filled in by *searching by rows.*

Appendix F

List of Symbols

Algebra

$\in$	$l \in L$; l is an element of L.
$\Re$	field of real numbers.
$\mathcal{C}$	field of complex numbers.
$\mathcal{J}$	ring of integers, namely, $\{\ldots, -1, 0, 1, 2, \ldots\}$.
$\underline{n}$	the set of integers $\{1, 2, \ldots, n\}$.
$\Re[q]$	ring of polynomials in one variable, q, with coefficients in the field $\Re$.
$\Re(q)$	ring of proper or strictly proper rational functions in q with coefficients in $\Re$.
L^n	set of n-tuples of elements belonging to the set L, (e.g., $\Re^n$, $\Re[q]^n$).
$L^{i \times j}$	set of $i \times j$ arrays of elements belonging to the set L or, equivalently, the set of $i \times j$ matrices with entries in the set L, (e.g., $\Re^{i \times j}$, $\Re[q]^{i \times j}$).
C_r^n	the number of different combinations of n different things, r at a time, without repetitions.

(see next page for symbols for matrices and polynomial matrices)

Matrices and Polynomial Matrices

L^T	the transpose of a matrix L.
L^b	the block–transpose of a block–matrix L.
L^{-1}	the inverse of a matrix L.
$n_r(L)$	the number of rows of a matrix L.
$n_c(L)$	the number of columns of a matrix L.
$\text{rank}(L)$	the rank of a matrix L.
$\det(L)$	the determinant of a matrix L.
$\text{Adj}(L)$	the adjoint of a matrix L.
$L.M$	the product of two conformable matrices L and M.
$\otimes$	$L \otimes M$ is the Kronecker product of matrices L and M.
$L(q)$	a polynomial matrix.
$\times$	$n_r \times n_c$ denotes the dimension of a n_r row by n_c column matrix.
$\partial[L(q)]$	the degree of the polynomial matrix $L(q)$.
$\partial_{ri}[L(q)]$	the ith row degree of the polynomial matrix $L(q)$.
$\partial_{cj}[L(q)]$	the jth column degree of the polynomial matrix $L(q)$.
$\mathcal{F}(q)$	the vector space $\mathcal{F}(q)$ (also called the linear space).
$P_c(q)$	the column structure matrix.
$E_c(q)$	the column residue matrix.
$P_r(q)$	the row structure matrix.
$E_r(q)$	the row residue matrix.
$\Gamma_c[L(q)]$	the highest column-degree coefficient matrix of $L(q)$.
$\Gamma_r[L(q)]$	the high row-degree coefficient matrix of $L(q)$.
Ls	the skeleton row block–matrix of a polynomial matrix $L(q)$, the column structure matrix of which is $P_c(q)$.
L^bs	the skeleton column block–matrix of a polynomial matrix $L(q)$, the row structure matrix of which is $P_r(q)$.
$\text{diag}[l_1, \ldots, l_n]$	a $n \times n$ diagonal matrix of diagonal entries $l_1, \ldots, l_n$.
$< L >_h$	the resultant of order h of a polynomial matrix $L(q)$.
$Z_k^{-1}(< L >_h)$	the pseudo-inverse of order k of a resultant $< L >_h$.
$TL_g(L)$	the lower-triangular Toeplitz block–matrix of order g formed from the coefficients of a polynomial matrix $L(q)$.
$TU_g(L)$	the upper-triangular Toeplitz block–matrix of order g formed from the coefficients of a polynomial matrix $L(q)$.
$C_m(L(q))$	the companion matrix of a monic polynomial matrix $L(q)$.

Linear Systems

t	time operator.
q	differential or forward difference operator.
z	the Z-transform operator.
s	the Laplacian transform operator.
$\mathcal{Z}\{f\}$	Z-transform of the function f.
$\mathcal{L}\{f\}$	Laplace transform of the function f.
n_{i}	the number of system inputs.
n_{o}	the number of system outputs.
n_{f}	the number of system outputs available for feedback.
$r(t)$	the reference inputs.
$u(t)$	the actuator (or error) inputs.
$c(t)$	the controlled outputs.
$c_{\mathrm{F}}(t)$	the controlled outputs available for feedback.
$G(q)$	the plant transfer function.
$G_{\mathrm{F}}(q)$	the modified plant transfer function.
$F(q)$	the transfer function of the output feedback compensator.
$R(q)$	the transfer function of the pre-compensator.
$H(q)$	the transfer function of a closed-loop system.
$S(A, B, C, E)$	a state-space system, where A, B, C and E are the plant, input, output and transmission matrices, respectively.
$\mathcal{OB}_j$	the observability matrix of order j.
$\mathcal{CA}_j$	the observability matrix of order j with the order of the block-elements reversed.
$\mathcal{CT}_j$	the controllability matrix of order j.
$\mathcal{AB}_j$	the controllability matrix of order j with the order of the block–elements reversed.
M_j	the matrix of Markov parameters of order j.
$\nu_1, \ldots, \nu_{n_{\mathrm{o}}}$	the set of column indices of a n_{o} column polynomial matrix, or the set of observability indices of a proper or strictly proper rational $n_{\mathrm{o}} \times n_{\mathrm{i}}$ transfer function.
$\mu_1, \ldots, \mu_{n_{\mathrm{i}}}$	the set of row indices of a n_{i} row polynomial matrix, or the set of controllability indices of a proper or strictly proper rational $n_{\mathrm{o}} \times n_{\mathrm{i}}$ transfer function.

Bibliography

[1] D. K. Anand, *Introduction to Control Systems.* London: Pergamon Press, 1984.

[2] A. N. Andry, E. Y. Shapiro and J. C. Chung, "Eigenstructure assignment for linear systems," *IEEE Trans. Aerospace and Electronic Systems*, vol. AES-19, pp. 711-729, September 1983.

[3] B. D. O. Anderson and J. B. Moore, *Optimal Filtering.* Prentice-Hall, 1979.

[4] H. Anton, *Elementary Linear Algebra.* New York: Wiley, 1977.

[5] H. Akaike, "Block Toeplitz matrix inversion," *SIAM Journ. Appl. Math*, vol. 24, no. 2, pp. 234-241, March 1973.

[6] M. Athans and T. L. Johnson, "On the design of optimal constrained dynamic compensation for linear constant systems," *IEEE Trans. Automat. Contr.*, vol. AC-15, pp. 658-660, 1970.

[7] S. Barnett, *Polynomials and Linear Control Systems.* New York: Van Nostrand Reinhold, 1972.

[8] S. Barnett, *Matrices in Control Theory.* New York: Van Nostrand Reinhold, 1972.

[9] S. Barnett, *Introduction to Mathematical Control Theory.* New York: Clarendon, 1975.

[10] M. Bellanger, *Digital Processing of Signals, Theory and Practice.* New York: Wiley-Interscience, 1986.

[11] W. A. Berger, R. J. Perry and H. H. Sun, "An algorithm for the assignment of system zeros," *Automatica*, vol. 27, no. 3, pp. 541-544, 1991.

[12] J. Belletrutti, "A note on pole assignment," *Control Systems Centre Report No. 207*, UMIST, Manchester, England.

[13] J. Bokor and L. Keviczky, "Structural properties and structure estimation of vector difference equations," *Int. J. Contr.* vol. 36, no. 3, pp. 461-475, 1982.

[14] F. M. Brasch, Jr. and J. B. Pearson, "Pole placement using dynamic compensators," *IEEE Trans. Automat. Contr.*, vol. AC-15, no. 1, pp. 34-43, February 1970.

[15] P. Brunovský, "A classification of linear multivariable systems," *Kybernetika*, vol. 6, pp. 173-188, 1970.

[16] F. M. Callier and C. A. Desoer, *Multivariable Feedback Systems.* Springer-Verlag, 1982.

[17] Chi-Tsong Chen, *Linear System Theory and Design.* Holt, Rinehart and Winston, 1984.

[18] C. T. Chen, *Joint Automatic Control Conference,* pp. 497, 1970.

[19] C. T. Chen and C. H. Hsu, *Joint Automatic Control Conference,* pp. 893, 1971.

[20] Chern-Lin Chen and Yuan-Yih Hsu, "Pole assignment using dynamic output feedback compensators," *Int. J. Contr.*, vol. 45, no. 6, pp. 1985-1994, 1987.

[21] D. Commenges and M. Monsion, "Fast inversion of triangular Toeplitz matrices," *IEEE Trans. Automat. Contr.*, vol. AC-29, pp. 250-251, March 1984.

[22] E. J. Davison, "On pole assignment in linear system with incomplete state feedback," *IEEE Trans. Automat. Contr.*, vol. AC-15, pp. 348-351, June 1970.

[23] E. J. Davison and S. H. Wang, "On pole assignment in linear multivariable systems using output feedback," *IEEE Trans. Automat. Contr.*, vol. AC-20, no. 4, pp. 516-518, August 1975.

[24] E. J. Davison and S. H. Wang, "Properties of linear time-invariant multivariable systems subject to arbitrary output and state feedback," *IEEE Trans. Automat. Contr.*, vol. AC-18, pp. 24-32, February 1973.

[25] C. A. Desoer and J.D. Schulman, "Zeros and poles of matrix transfer functions and their dynamical interpretation," *IEEE Trans. Circuits Syst.*, vol. CAS-21, pp. 3-8, January 1974.

[26] C. A. Desoer, R. Liu, J. Murray and R. Saeks, "Feedback systems design: the fractional representation approach to analysis and synthesis," *IEEE Trans. Automat. Contr.*, vol. AC-25, no. 3, pp. 401-412, June 1980.

[27] B. W. Dickinson, M. Morf and T. Kailath, "A minimal realization algorithm for matrix sequences," *IEEE Trans. Automat. Contr.*, vol. AC-19, pp. 31-38, February 1974.

[28] A. Dickman and R. Sivan, "On the robustness of multivariable linear feedback systems," *IEEE Trans. Automat. Contr.*, vol. AC-30, no. 4, April 1985.

[29] C. Y. Ding and J. B. Pearson, "Compensator design for multivariable linear systems," *IEEE Trans. Automat. Contr.*, vol. AC-14, pp. 130-134, April 1969.

[30] J. Dongarra, "LINPACK Users' Guide," *SIAM*, Philadelphia, 1979.

[31] P. M. Van Dooren, "The generalized eigenstructure problem in linear system theory," *IEEE Trans. Automat. Contr.*, vol. AC-26, no. 1, pp. 111-128, February 1981.

[32] J. M. Edmunds, I. Postlethwaite and A. G. J. Macfarlane, "Principal gains and principal phases in the analysis of linear multivariable feedback systems," *IEEE Trans. Automat. Contr.*, vol. AC-26, no. 1, pp. 32-46, February 1981.

[33] H. Elliot, W. A. Wolovich and M. Das,"Arbitrary adaptive pole placement for linear multivariable systems," *IEEE Trans. Automat. Contr.*, vol. AC-29, pp. 221-228, March 1984.

[34] E. Emre and L. Silverman, "The equation $XR + QY = \Phi$: a characterization of solutions," *SIAM J. Contr. Optimiz.*, vol. 19, no. 1, pp. 33-38, 1981.

[35] E. Emre, "Nonsingular factors of polynomial matrices and (A, B)-invariant subspaces," *SIAM J. Contr. Optimiz.*, vol. 18, no. 3, pp. 288-296, May 1980.

[36] M. Fahmy and J. O'Reilly, "Multistage parametric eigenstructure assignment by output feedback," *Int. J. Contr.*, vol. 48, pp. 97-116, 1988.

[37] F. Fallside, Ed., *Control System Design by Pole-Zero Assignment.* Academic Press, 1977.

[38] F. Fallside and H. Seraji, "Pole-shifting procedure for multivariable systems using output feedback," *Proc. IEE*, vol. 25, pp. 483-490, 1977.

[39] J. Feinstein, and Y. Bar-ness, "The solution of the matrix polynomial equation $A(s)X(s) + B(s)Y(s) = C(s)$," *IEEE Trans. Automat. Contr.*, vol. AC-29, pp. 75-77, January 1984.

[40] G. D. Forney, Jr., "Minimal bases of rational vector spaces with applications to multivariable linear systems," *SIAM J. Contr.*, vol. 13, no. 3, pp. 493-520, May 1975.

[41] A. J. Fossard, *Multivariable System Control.* Netherlands: North-Holland Publishing Co., 1977

[42] L. V. Foster, "A practical solution to the minimal design problem," *IEEE Trans. Automat. Contr.*, vol. AC-24, pp. 449-454, 1979.

[43] G. F. Franklin and J. D. Powell, *Digital Control of Dynamic Systems.*, Addison Wesley, January 1980.

[44] P. A. Fuhrmann, "Algebraic system theory: an analyst's point of view," *J. Franklin Inst.*, vol. 301, no. 6, pp. 521-540, June 1976.

[45] M. J. Gibbard and A. P. Papliński, "A new class of computational methods and associated methodology for the design of digital controllers for multi-input, multi-output linear systems," Technical report CONT 86/01, University of Adelaide, January 1986.

[46] I. Gohberg, P. Lancaster and L. Rodman, *Matrix Polynomials.*, Academic Press, 1982.

[47] T. N. E. Greville, "The pseudo-inverse of a rectangular or singular matrix and its application to the solution of systems of linear equations," *SIAM Rev.*, vol. 1, no. 1, pp. 38-43, January 1959.

[48] M. R. Gevers, "ARMA models, their Kronecker indices and their McMillan degree," *Int. J. Contr.*, vol. 43, no. 6, pp. 1745-1761, 1986.

[49] R. Guidorzi, "Canonical structures in the identification of multivariable systems," *Automatica*, vol. 11, pp. 361-374, 1975.

[50] A. N. Gündeş and C. A. Desoer, *Algebraic Theory of Linear Feedback Systems with Full and Decentralized Compensators.* Lecture Notes in Control and Information Sciences, Springer–Verlag, 1990.

[51] C. L. Gustafson and C. A. Desoer, "A CAD methodology for linear multivariable feedback systems based on algebraic theory," *Int. J. Contr.*, vol. 41, no. 3, pp. 653-675, 1985.

[52] J. Hammer, "Linear dynamic output feedback: invariants and stability," *IEEE Trans. Automat. Contr.*, vol. AC-28, no. 4, April 1983.

[53] Integrated Systems Inc., 2500 Mission College Blvd., Santa Clara, CA 95054–1215, U.S.A. (MATRIX$_X$).

[54] I. S. Iohvidov, *Hankel and Toeplitz Matrices and Forms. Algebraic Theory.* Birkhauser Verlag, 1982.

[55] A. Jameson, "Design of single input systems for specified roots using output feedback," *IEEE Trans. Automat. Contr.*, vol. AC-15, pp. 345-348, June 1970.

[56] T. Kailath, *Modern Signal Processing.* New York: Springer-Verlag, 1985.

[57] T. Kailath, *Linear Systems.* Englewood Cliffs, NJ: Prentice-Hall, 1980.

[58] T. Kailath and S. Kung, "Fast projection methods for minimal design problems in linear system theory,"*Automatica*, vol. 16, pp. 399-403, July 1980.

[59] R. E. Kalman, "Kronecker invariants and feedback", In *Ordinary Differential Equations.* Academic Press, New York, pp. 459-471, 1972.

[60] R. E. Kalman and R. S. Bucy, "New results in linear filtering and prediction theory," *IEEE Trans. ASME, J. Basic Eng.*, Ser. D., pp. 95-108, 1961.

[61] H. Kimura, " Pole assignment by gain output feedback," *IEEE Trans. Automat. Contr.*, vol. AC-20, no. 4, pp. 509-516, August 1975.

[62] H. Kimura, "A further result on the problem of pole assignment by output feedback," *IEEE Trans. Automat. Contr.*, vol. AC-22, pp. 458-463, June 1977.

[63] T. G. Koussiouris, "Controllability indices of a system, minimal indices of its transfer function matrix, and their relations," *Int. J. Contr.*, vol. 34, no. 3, pp. 613-622, 1981.

[64] S. Kung, T. Kailath and M. Morf, "Fast and stable algorithms for minimal design problems," *Proc. 4th IFAC Int. Symp. Multivariable Technol. Syst.*, D. P. Atherton, Ed. London: Pergamon, pp. 97-104, 1977.

[65] S. Kung, T. Kailath and M. Morf, "A generalised resultant matrix for polynomial matrices", *Proc. 1976 IEEE Conference on Decision and Control*, pp. 892-895.

[66] V. Kučera, "Stochastic multivariable control: a polynomial equation approach," *IEEE Trans. Automat. Contr.*, vol. AC-25, pp. 913-919, 1980.

[67] G. Lee, D. Jordan and M. Sohrwardy, "A pole assignment algorithm for multivariable control systems," *IEEE Trans. Automat. Contr.*, vol. AC-24, no. 2, pp. 357- , April 1979.

[68] J. R. Leigh, *Applied Control Theory.* Exeter, Short Run, 1982.

[69] N. A. Lehtomaki, "Practical robustness measures in multivariable control system analysis," Ph.D. Thesis, Massachusetts Institute of Technology, Cambridge, May 1981.

[70] D. G. Luenberger, "Canonical forms for linear multivariable systems," *IEEE Trans. Automat. Contr.*, vol. AC-12, pp. 290-293, June 1967.

[71] D. G. Luenberger, "An introduction to observers," *IEEE Trans. Automat. Contr.*, vol. AC-16, pp. 596-602, December 1971.

[72] D. G. Luenberger, "Observers for multivariable systems," *IEEE Trans. Automat. Contr.*, vol. AC-11, pp. 190-197, April 1966.

[73] The MathWorks Inc., Cochituate Place, 24 Prime Park Way, Natick, MA 01760, U.S.A. (PRO–MATLAB).

[74] C. C. Macduffee, *The Theory of Matrices.*, New York, Chelsea, 1956.

[75] A. G. J. Macfarlane, *Complex Variable Methods for Linear Multivariable Feedback Systems.* London: Taylor & Francis, 1980.

[76] P. Misra and R. V. Patel, "Transmission zero assignment in linear multivariable systems," *IEEE Conference on Decision and Control*, Austin, Texas, 1988.

[77] C. Moler and C. Van Load, "Nineteen dubious ways to compute the exponential of a matrix," *SIAM Rev.*, vol. 20, pp. 801-836, October 1978.

[78] C. Moler, *MATLAB Users' Guide*, Dept. of Computer Science, University of New Mexico, May 1981.

[79] B. C. Moore and L. M. Silverman, "Model matching by state feedback and dynamic compensation," *IEEE Trans. Automat. Contr.*, vol. AC-17, no. 4, pp. 491-496, August 1972.

[80] A. S. Morse, "Minimal solutions to transfer matrix equations," *IEEE Trans. Automat. Contr.*, vol. AC-21, pp.131-133, 1976.

[81] N. Munro, *Modern Approaches to Control System Design.* London: A. Wheaton & Co., 1979.

[82] N. Munro and A. I. Vardulakis, "Pole-shifting using output feedback," *Int. J. Contr.*, vol. 18, pp. 1267-1273, 1973.

[83] N. Munro and S. Novin-Hirbod, "Pole assignment using full rank output feedback compensators," *Int. J. System Science*, vol. 10, no. 3, pp. 285-306, 1979.

[84] N. K. Nichols, J. Kautsky and P. Van Dooren, "Robust pole assignment in linear state feedback," *Int. J. Contr.*, vol. 41, pp. 1129-1155, 1985.

[85] N. K. Nichols, "On computational algorithms for pole assignment," *IEEE Trans. Automat. Contr.*, vol. AC-31, no. 7, pp. 643-645, July 1986.

[86] A. P. Papliński, "Matrix multiplication and division of polynomials" *IEEE Proc.*, vol. 132, Pt. D, no. 3, pp. 95-99, May 1985.

[87] A. P. Papliński, "Elementary operations with block matrices," Technical Report CONT 85/01, University of Adelaide, July 1985.

[88] A. P. Papliński, "Discretizing a multivariable continuous system with pure intergration," *IEEE Trans. Automat. Contr.*, vol. AC-30, no. 5, pp. 510-512, May 1985.

[89] A. P. Papliński and M. J. Gibbard, "On constraints in pole assignment by static feedback," *Int. J. Contr.*, vol. 42, no. 5, pp. 1099-1113, November 1985.

[90] A. P. Papliński, "Transforming a polynomial matrix into a monic or comonic form," Technical report CONT 87/01, University of Adelaide, April 1987.

[91] R. V. Patel, "Design of dynamic compensators for pole assignment," *Int. J. Systems Sci.*, vol. 7, no. 2, pp. 207-224, 1976.

[92] R. V. Patel and N. Munro, *Multivariable System Theory and Design.* Oxford: Pergamon Press, 1982.

[93] R. V. Patel, "Computation of matrix fraction descriptions of linear time-invariant systems," *IEEE Trans. Automat. Contr.*, vol. AC-26, pp. 148-161, February 1981.

[94] R. V. Patel, "On output feedback pole assignability," *Int. J. Contr.*, vol. 20, pp. 955-959, 1974.

[95] J. B. Pearson and C. Y. Ding, "Compensator design for multivariable linear systems," *IEEE Trans. Automat. Contr.*, vol. AC-14, no. 2, pp. 130-134, April 1969.

[96] P. H. R. Petkov, N. D. Christov and M. M. Kostantinov, "A computational algorithm for pole assignment of linear multi-input systems," *IEEE Trans. Automat. Contr.*, vol. AC-31, no. 11, pp. 1044-1047, November 1986.

[97] R. A. Pine, R. A. Walker and N. K. Gupta, "$MATRIX_X/PC$: modeling, simulation, and optimization on personal computers," *Proc. IEEE*, vol. 73, no. 12, pp. 1817-1825, December 1985.

[98] V. M. Popov, "Invariant description of linear, time-invariant controllable systems," *SIAM J. Contr.*, vol. 10, pp. 252-264, May 1972.

[99] V. M. Popov, *Some Properties of the Control Systems with Irreducible Matrix Transfer Functions.* New York: Springer, 1970, pp. 250-261

[100] J. Rissanen, "Basis of invariants and canonical forms for linear dynamic systems," *Automatica*, vol. 10, pp. 175-182, 1974.

[101] H. H. Rosenbrock, *Computer-Aided Control System Design.* Academic Press, 1974.

[102] H. H. Rosenbrock, *State-Space and Multivariable Theory.* New York: Wiley-Interscience, 1970.

[103] H. H. Rosenbrock and G. E. Hayton, "The general problem of pole assignment," *Int. J. Contr.*, vol. 27, pp. 837-852, 1978.

[104] H. H. Rosenbrock and G. E. Hayton, "Dynamical indices of a transfer function matrix," *Int. J. Contr.*, vol. 20, no. 2, pp. 177-189, 1974.

[105] M. K. Sain, "A free-modular algorithm for minimal design of linear multivariable systems," 1975 IFAC *Congr. Preprints*, paper 9.1, part 1B.

[106] O. A. Sebakhy and N. N. Sorial, "Optimization of linear multivariable systems with prespecified closed-loop eigenvalues," *IEEE Trans. Automat. Contr.*, vol. AC-24, no. 2, pp. 355-357, April 1979.

[107] P. K. Sinha, *Multivariable Control, An Introduction.* New York: Dekker, 1974.

[108] B. Sridhar and D. P. Lindorff, "Pole-placement with constant gain output feedback," *Int. J. Contr.*, vol. 18, pp. 993-1003, Nov. 1973.

[109] B. T. Smith *et al.*, "Matrix eigensystem routines-EISPACK guide," 2nd ed., in *Lecture Notes in Computer Science*, vol. 6. Berlin: Springer-Verlag, 1976.

[110] K. M. Sobel and E. Y. Shapiro, "Application of eigenstructure assignment to flight control design," *Journal of Guidance, Control and Dynamics*, vol. 10, pp. 73-81, Jan.-Feb. 1987.

[111] Y. T. Tsay, L.–S. Shieh and S. Barnett, *Structural Analysis and Design of Multivariable Systems: an Algebraic Approach.* Lecture Notes in Control and Information Sciences, Springer–Verlag, 1988.

[112] A. I. G. Vardulakis, "On the structure of the bases of all possible controllability subspaces $\Re$ of a controllable pair [A, B] in canonical form," *Int. J. Contr.*, vol. 28, no. 3, pp. 393-409, 1978.

[113] G. C. Verghese, "Infinite-frequency behaviour in generalized dynamical systems." Ph.D. Thesis, Electrical Engg. Dept., Stanford University, Stanford, CA., 1978.

[114] M. Vidyasagar,*Control System Synthesis: A Factorization Approach.* The MIT Press, 1985.

[115] S. H. Wang and E. J. Davison, "A minimization algorithm for the design of multivariable systems", *IEEE Trans. Automat. Contr.*, vol. AC-18, pp. 31-38, February 1973.

[116] B. A. White, "Eigenstructure assignment by output feedback," *Int. J. Contr.*, vol. 53, no. 6, pp. 1413-1429, 1991.

[117] J. C. Willems and W. H. Hesselink, "Generic properties of the pole placement problem," *7th World Congress of IFAC*, Helsinki, 1978.

[118] W. A. Wolovich, *Linear Multivariable Systems.* New York: Springer-Verlag, 1974.

[119] W. A. Wolovich, "The determinations of state-space representations for linear multivariable systems," *Automatica*, vol. 9, no. 1, pp. 97-106, 1973.

[120] W. A. Wolovich and P. L. Falb, "On the structure of multivariable systems," *SIAM J. Contr.*, vol. 7, pp. 437-451, 1969.

[121] W. A. Wolovich,"A division algorithm for polynomial matrices," *IEEE Trans. Automat. Contr.*, vol. AC-29, no. 7, pp. 656-658, July 1984.

[122] W. A. Wolovich,"The determination of state-space representations for linear multivariable systems," presented at the IFAC Symp., Dusseldorf, Germany, October 1971.

[123] W. A. Wolovich and P. L. Falb, "Invariants and canonical forms under dynamic compensation," *SIAM J. Contr. Optimiz.*, vol. 14, pp. 996-1008, November 1976.

[124] W. M. Wonham,"On pole assignment in multi-input, controllable linear systems," *IEEE Trans. Automat. Contr.*, vol. AC-12, pp. 660-665, 1967.

[125] W. M. Wonham and A. S. Morse,"Feedback invariants of linear multivariable systems," *Automatica*, vol. 8, pp. 93-100, 1972.

[126] S. H. Zak, "On the polynomial matrix equation $AX + YB = C$," *IEEE Trans. Automat. Contr.*, vol. AC-30, no. 12, pp. 1240-1242, December 1985.

[127] S. Zohar,"The solution of a Toeplitz set of linear equations," *J. ACM*, vol. 21, no. 2, pp. 273-276, April 1974.

Index

Lecture Notes in Control and Information Sciences

Edited by M. Thoma and A. Wyner

Vol. 117: K.J. Hunt
Stochastic Optimal Control Theory
with Application in Self-Tuning Control
X, 308 pages, 1989.

Vol. 118: L. Dai
Singular Control Systems
IX, 332 pages, 1989

Vol. 119: T. Başar, P. Bernhard
Differential Games and Applications
VII, 201 pages, 1989

Vol. 120: L. Trave, A. Titli, A. M. Tarras
Large Scale Systems:
Decentralization, Structure Constraints
and Fixed Modes
XIV, 384 pages, 1989

Vol. 121: A. Blaquière (Editor)
Modeling and Control of Systems
in Engineering, Quantum Mechanics,
Economics and Biosciences
Proceedings of the Bellman Continuum
Workshop 1988, June 13–14, Sophia Antipolis, France
XXVI, 519 pages, 1989

Vol. 122: J. Descusse, M. Fliess, A. Isidori,
D. Leborgne (Eds.)
New Trends in Nonlinear Control Theory
Proceedings of an International
Conference on Nonlinear Systems,
Nantes, France, June 13–17, 1988
VIII, 528 pages, 1989

Vol. 123: C. W. de Silva, A. G. J. MacFarlane
Knowledge-Based Control with
Application to Robots
X, 196 pages, 1989

Vol. 124: A. A. Bahnasawi, M. S. Mahmoud
Control of Partially-Known
Dynamical Systems
XI, 228 pages, 1989

Vol. 125: J. Simon (Ed.)
Control of Boundaries and Stabilization
Proceedings of the IFIP WG 7.2 Conference
Clermont Ferrand, France, June 20–23, 1988
IX, 266 pages, 1989

Vol. 126: N. Christopeit, K. Helmes
M. Kohlmann (Eds.)
Stochastic Differential Systems
Proceedings of the 4th Bad Honnef Conference
June 20–24, 1988
IX, 342 pages, 1989

Vol. 127: C. Heij
Deterministic Identification
of Dynamical Systems
VI, 292 pages, 1989

Vol. 128: G. Einarsson, T. Ericson,
I. Ingemarsson, R. Johannesson,
K. Zigangirov, C.-E. Sundberg
Topics in Coding Theory
VII, 176 pages, 1989

Vol. 129: W. A. Porter, S. C. Kak (Eds.)
Advances in Communications and
Signal Processing
VI, 376 pages, 1989.

Vol. 130: W. A. Porter, S. C. Kak,
J. L. Aravena (Eds.)
Advances in Computing and Control
VI, 367 pages, 1989

Vol. 131: S. M. Joshi
Control of Large Flexible Space Structures
IX, 196 pages, 1989.

Vol. 132: W.-Y. Ng
Interactive Multi-Objective Programming
as a Framework for Computer-Aided Control
System Design
XV, 182 pages, 1989.

Vol. 133: R. P. Leland
Stochastic Models for Laser Propagation
in Atmospheric Turbulence
VII, 145 pages, 1989.

Vol. 134: X. J. Zhang
Auxiliary Signal Design in Fault
Detection and Diagnosis
XII, 213 pages, 1989

Vol. 135: H. Nijmeijer, J. M. Schumacher (Eds.)
Three Decades of Mathematical System Theory
A Collection of Surveys at the Occasion of the
50th Birthday of Jan C. Willems
VI, 562 pages, 1989

Vol. 136: J. Zabczyk (Ed.)
Stochastic Systems and Optimization
Proceedings of the 6th IFIP WG 7.1
Working Conference,
Warsaw, Poland, September 12–16, 1988
VI, 374 pages. 1989

Lecture Notes in Control and Information Sciences

Edited by M. Thoma and A. Wyner

Vol. 137: S. L. Shah, G. Dumont (Eds.)
Adaptive Control Strategies for
Industrial Use
Proceedings of a Workshop
Kananaskis, Canada, 1988
VI, 360 pages. 1989

Vol. 138: D. C. McFarlane, K. Glover
Robust Controller Design Using
Normalized Coprime Factor
Plant Descriptions
X, 206 pages. 1990

Vol. 139: V. Hayward, O. Khatib (Eds.)
Experimental Robotics I
The First International Symposium
Montreal, June 19-21, 1989
XVII, 613 pages. 1990

Vol. 140: Z. Gajic, D. Petkovski,
X. Shen
Singularly Perturbed and Weakly
Coupled Linear Control Systems
A Recursive Approach
VII, 202 pages. 1990

Vol. 141: S. Gutman
Root Clustering in Parameter Space
VIII, 153 pages. 1990

Vol. 142: A. N. Gündeş, C. A. Desoer
Algebraic Theory of Linear
Feedback Systems with Full and
Decentralized Compensators
V, 176 pages. 1990

Vol. 143: H.-J. Sebastian, K. Tammer (Eds.)
System Modelling and Optimization
Proceedings of the 14th IFIP Conference
Leipzig, GDR, July 3-7, 1989
X, 960 pages. 1990

Vol. 144: A. Bensoussan, J. L. Lions (Eds.)
Analysis and Optimization of Systems
Proceedings of the 9th International
Conference, Antibes, June 12-15, 1990
XII, 992 pages. 1990

Vol. 145: M. B. Subrahmanyam
Optimal Control with a Worst-Case
Performance Criterion and Applications
X, 133 pages. 1990

Vol. 146: D. Mustafa, K. Glover
Minimum Entropy H_∞ Control
IX, 144 pages. 1990

Vol. 147: J. P. Zolésio (Ed.)
Stabilization of Flexible Structures
Third Working Conference,
Montpellier, France, January 1989
V, 327 pages, 1990

Vol. 148: In preparation

Vol. 149: K. H. Hoffmann, W. Krabs (Eds.)
Optimal Control of Partial
Differential Equations
Proceedings of the IFIP WG 7.2 International
Conference, Irsee, April 9-12, 1990
VI, 245 pages. 1991

Vol. 150: L. C. G. J. M. Habets
Robust Stabilization
in the Gap-topology
IX, 126 pages. 1991

Vol. 151: J. M. Skowronski,
H. Flashner, R. S. Guttalu (Eds.)
Mechanics and Control
Proceedings of the 3rd Workshop on
Control Mechanics
in Honor of the 65th Birthday of
George Leitmann
January 22-24, 1990, University of Southern
California
IV, 497 pages. 1991

Vol. 152: J. D. Aplevich
Implicit Linear Systems
XI, 176 pages. 1991

Vol. 153: O. Hájek
Control Theory in the Plane
X, 272 pages. 1991

Vol. 154: A. Kurzhanski, I. Lasiecka (Eds.)
Modelling and Inverse Problems of Control
for Distributed Parameter Systems
Proceedings of IFIP (W.G. 7.2)-IIASA Conference
Laxenburg, Austria, July 24-28, 1989
VII, 179 pages. 1991

Vol. 155: M. Bouvet, G. Bienvenu (Eds.)
High-Resolution Methods in Underwater
Acoustics
V, 249 pages. 1991

Lecture Notes in Control and Information Sciences

Edited by M. Thoma and A. Wyner

Vol. 156: R. P. Hämäläinen, H. K. Ehtamo (Eds.)
Differential Games –
Developments in Modelling and Computation
Proceedings of the Fourth International Symposium
on Differential Games and Applications
August 9-10, 1990, Helsinki University of Technology,
Finland
XIII, 292 pages. 1991

Vol. 157: R. P. Hämäläinen, H. K. Ehtamo (Eds.)
Dynamic Games in Economic Analysis
Proceedings of the Fourth International Symposium
on Differential Games and Applications
August 9-10, 1990, Helsinki University of Technology,
Finland
XIII, 311 pages. 1991

Vol. 158: K. Warwick, M. Kárný,
A. Halousková (Eds.)
Advanced Methods in Adaptive Control
for Industrial Applications
X, 331 pages. 1991

Vol. 159: X. Li, J. Yong (Eds.)
Control Theory of Distributed Parameter Systems
and Applications
Proceedings of the IFIP WG 7.2 Working Conference,
Shanghai, China, May 6–9, 1990
VIII, 219 pages. 1991

Vol. 160: P. V. Kokotović (Ed.)
Foundations of Adaptive Control
IX, 525 pages. 1991

Vol. 161: L. Gerencsér, P. E. Caines (Eds.)
Topics in Stochastic Systems:
Modelling, Estimation and Adaptive Control
IV, 401 pages, 1991

Vol. 162: C. Canudas de Wit (Ed.)
Advanced Robot Control
IX, 314 pages, 1991

Vol. 163: V. L. Mehrmann
The Autonomous Linear Quadratic
Control Problem
Theory and Numerical Solution
VI, 177 pages, 1991

Vol. 164: I. Lasiecka, R. Triggiani
Differential and Algebraic Riccati Equations
with Application to Boundary/Point
Control Problems: Continuous Theory
and Approximation Theory
XI, 160 pages, 1991

Vol. 165: G. Jacob, F. Lamnabhi-Lagarrigue (Eds.)
Algebraic Computing in Control
Proceedings of the First European Conference
Paris, March 13–15, 1991
IX, 385 pages. 1991

Vol. 166: L. L. M. van der Wegen
Local Disturbance Decoupling
with Stability for Nonlinear Systems
V, 135 pages, 1991

Vol. 167: M. Rao
Integrated System for Intelligent Control
VIII, 133 pages. 1992

Vol. 168: P. Dorato, L. Fortuna, G. Muscato
Robust Control for Unstructured Perturbations –
An Introduction
VI, 117 pages. 1992

Vol. 169: In preparation

Vol. 170: J. M. Skowronski, H. Flashner,
R. S. Guttalu (Eds.)
Mechanics and Control
Proceedings of the 4th Workshop
on Control Mechanics, January 21–23, 1991
University of Southern California, USA
IV, 301 pages. 1992

Vol. 171: P. Stefanidis, A. P. Papliński,
M. J. Gibbard
Numerical Operations with Polynomial Matrices
Application to Multi-Variable Dynamic
Compensator Design
VIII, 205 pages. 1992

GPSR Compliance

The European Union's (EU) General Product Safety Regulation (GPSR) is a set of rules that requires consumer products to be safe and our obligations to ensure this.

If you have any concerns about our products, you can contact us on ProductSafety@springernature.com

In case Publisher is established outside the EU, the EU authorized representative is:

Springer Nature Customer Service Center GmbH
Europaplatz 3
69115 Heidelberg, Germany

Batch number: 10704763

Printed by Printforce, the Netherlands